BIOCHEMISTRY RESEARCH TRENDS

ETHANOL

PRODUCTION, CELLULAR MECHANISMS AND HEALTH IMPACT

BIOCHEMISTRY RESEARCH TRENDS

Additional books in this series can be found on Nova's website
under the Series tab.

Additional e-books in this series can be found on Nova's website
under the e-book tab.

BIOCHEMISTRY RESEARCH TRENDS

ETHANOL

PRODUCTION, CELLULAR MECHANISMS AND HEALTH IMPACT

ROY I. HENRY
AND
BILLY P. WOODS
EDITORS

NOVA BIOMEDICAL

New York

For permission to use material from this book please contact us:
Telephone 631-231-7269; Fax 631-231-8175
Web Site: http://www.novapublishers.com

Additional color graphics may be available in the e-book version of this book.

Library of Congress Cataloging-in-Publication Data

Ethanol : production, cellular mechanisms, and health impact / [edited by] Roy I. Henry and Billy P. Woods.
pages cm
Includes bibliographical references and index.
ISBN 978-1-62257-297-7 (hardcover)
1. Ethanol as fuel. 2. Ethanol. I. Henry, Roy I., editor of compilation. II. Woods, Billy P., editor of compilation.
TP339.E85 2012
662.6'692--dc23

2012021401

Published by Nova Science Publishers, Inc. † New York

CONTENTS

PREFACE

The fermentation of sugar into ethanol is one of the earliest biotechnologies employed by humanity. In modern times, ethanol intended for industrial use is also produced from ethylene. Ethanol has widespread use as a solvent of substances intended for human contact or consumption, including scents, flavorings, colorings, medicines, and as a biofuel. This book discusses ethanol production, cellular mechanisms and its health impacts. Topics include the production of ethanol from synthesis gas in flowed vapor-solid catalytic systems; producing hydrogen from ethanol reforming processes through inorganic membrane reactors; the potential role of epigenetic factors in the etiology of fetal alcohol spectrum disorder; ethanol investment decisions; cellulosic ethanol production through bioconversion of lignocellulosic biomass; and the effects of withdrawal from ethanol use on emotional learning.

Chapter 1 - The worldwide bioethanol demand is rapidly growing and the traditional sources are no longer available in large quantities due to their primary value and use as food. For this reason, other raw materials must be considered and improved microorganisms that increase the fuel ethanol productivity are required. The research associated with the implementation of new raw materials and the search for new fermentative bio-agents is constant. The profitability of the bioethanol production process is highly related to the type of feedstock involved and the used microorganism. One of the main results obtained from research on the bioethanol production is a major knowledge of the fermentation process. This feature has led to many technological advances increasing the potential for processes under no conventional conditions. Nevertheless, these advances have still great limitations and low yielding that impede their scaling up to industrial levels.

The selection of the proper microorganism and feedstock for ethanol fermentation processes should comply not only with the characteristics that maximize the productivity of bioethanol, but also with other important constraints such as no involvement in food security, reduction of undesired by- products and pollutant compounds. The large scale production of bioethanol currently makes use of various microorganisms and raw materials, but a large amount of this biofuel is produced by the fermentation of starch or sugar compounds. New developments include several types of lignocellulosic compounds, such as agroindustrial wastes, grasslands and domestic wastes. The traditional fermenting microorganisms are not very efficient degrading these compounds thus causing the implementation of different pretreatment techniques. Also, no conventional microorganisms capable of utilizing different substrates, like those isolated from natural environments, adapted to different conditions or genetically modified, must be considered.

This chapter is a review of the main raw materials and microorganisms used for the production of bioethanol at industrial and pilot scales, the metabolic routes and cellular mechanisms that govern the conversion of substrates, the requirements that must be met for a cost effective fermentation and the main perspectives related to the bioethanol production process.

Chapter 2 - A chemical process using synthesis gas (so-called as syngas, a mixed gas of CO and H_2) as a platform can produce ethanol from all kinds of organic resources (such as biomass, coal, and so on). The process includes the first gasification step for producing syngas from organic resources and the second liquefaction step for converting syngas to ethanol. The gasification is a thermochemical reaction and the liquefaction is a catalytic reaction. Thus the efficiency of the total process is determined by the performance of the catalysts for converting syngas to ethanol. Three types of catalysts (Cu-based, Rh-based, and MoS-based) for the synthesis of alcohols (including ethanol) were investigated from syngas in this study. Firstly, a highly active Cu-based catalyst Cs-Ni-Cu/$Ce_{0.8}Zr_{0.2}O_2$ was developed for producing mixed alcohols from syngas at a reaction temperature of 573 K under a reaction pressure of 3.0 MPa. Both the CO conversion and the selectivity for alcohols over Cs-Ni-Cu/$Ce_{0.8}Zr_{0.2}O_2$ were higher than those over the reported Cu-based catalysts. The selectivity for alcohols was very high (>80%) but the fraction of ethanol in total alcohols was not high over Cs-Ni-Cu/$Ce_{0.8}Zr_{0.2}O_2$ because it mainly formed methanol and iso-butanol. Secondly, developed was a highly active Rh-based catalyst Li-Fe-Rh/$Ce_{0.8}Zr_{0.2}O_2$ for producing ethanol from syngas at 548 K under 2.4 MPa. Both the CO conversion and the selectivity for ethanol over Li-Fe-Rh/$Ce_{0.8}Zr_{0.2}O_2$ were higher than those over the reported Rh-based catalysts. Li-Fe-Rh/$Ce_{0.8}Zr_{0.2}O_2$ selectively formed ethanol among various alcohols although the selectivity for by-product methane was relatively high (>20%). Li-Fe-Rh/$Ce_{0.8}Zr_{0.2}O_2$ had a good catalytic stability in a clean syngas but lost activity gradually in a syngas containing 10 ppm H_2S due to forming inactive RhS phase. Thirdly, developed was a highly active Mo-based catalyst K-Fe-Ni-MoS/Al_{13}-Mont for producing mixed alcohols from syngas at 543 K under 6 MPa. Both the CO conversion and the selectivity for ethanol over K-Fe-Ni-MoS/Al_{13}-Mont were higher than those over the reported MoS-based catalysts. K-Fe-Ni-MoS/Al_{13}-Mont formed mixed alcohols as the main oxygenate compounds. The fraction of ethanol in the product over K-Fe-Ni-MoS/Al_{13}-Mont was higher than that over Cs-Ni-Cu/$Ce_{0.8}Zr_{0.2}O_2$ but was lower than that over Li-Fe-Rh/$Ce_{0.8}Zr_{0.2}O_2$. K-Fe-Ni-MoS/Al_{13}-Mont lost activity gradually in a clean syngas due to losing sulfur from catalyst surface during the reaction but it had a good catalytic stability in a syngas containing 10 ppm H_2S.

Chapter 3 - Today, hydrogen is considered as the most viable energy carrier useful for many applications and, particularly, for fuel cells supplying. Taking into account the environmental contamination caused by the exploitation of derived of fossil fuels, hydrogen production from ethanol coming from biomass as a renewable source could constitute an interesting and strategic opportunity. This chapter describes the possibility of producing hydrogen from ethanol reforming processes through inorganic membrane reactors. The latter are proposed and described because useful to perform reactions limited by the equilibrium conversion because of the hydrogen selective separation capability of the inorganic membranes. Both benefits and drawbacks of the inorganic membrane reactor technology are examined and, when applied to reforming reactions, their performances such as hydrogen yield, hydrogen recovery, ethanol conversion and so on are emphasized and qualitatively compared to those of the conventional systems. Furthermore, a short overview from the open specialized literature on the most representative scientific results concerning ethanol reforming processes performed by inorganic membrane reactor technology is also given and discussed.

Chapter 4 - Alcohol is recognized as a classic teratogen capable of inducing a wide range of developmental abnormalities. Exposure to alcohol in utero is the major attributable cause of fetal alcohol spectrum disorder (FASD), which is characterized by craniofacial anomalies and neurodevelopmental impairments, such as cognitive and learning disabilities and behavioral disorders. FASD is recognized as the leading nongenetic cause of mental retardation. Recent studies indicate that epigenetic regulatory mechanisms which include DNA methylation, histone modifications and noncoding RNA expression, play a significant role in FASD development. Aberrant epigenetic alterations are known to disrupt key developmental events, especially the central nervous system. This review, examined the potential role of epigenetic factors in the etiology of FASD.

Chapter 5 - Recently the support of biofuel production has been a politically sensitive topic. Several government policies actively promote ethanol production via tax incentives and mandates. Additionally, politicians have pushed for ethanol as an environmentally friendly alternative to imported oil, as well as a way to boost farm profits and improve rural livelihoods. All these factors have coincided with investment in the industry in recent years. Even when excluding the U.S., which was the country with the largest fuel-ethanol production in 2009, the fuel-ethanol industry has been growing rapidly in the rest of the world (ROW). Ethanol-producing countries in the ROW include Brazil, Canada, China, and Thailand, as well as countries in Europe. There are 191 fuel-ethanol plants in the ROW, which is a little more than in the U.S., and 82% of them were built after 2005.

This chapter reviews some of the papers I have written analyzing what factors affect the decision to invest in building new ethanol plants using a dynamic structural econometric model of the investment timing game. The results of my research will help determine which policies and factors can promote fuel-ethanol industry development.

In Lin and Thome (2012), an estimate of a model of the investment timing game in corn ethanol plants in the United States. This model follows my previous work estimating a structural econometric model of the multi-stage dynamic investment timing game in offshore petroleum production. In Lin and Yi (2012a,b), a model of the investment timing game in ethanol plants worldwide that allows for the choice among different feedstocks.

According to results found , in the United States, competition between plants is enough to deter local investments. Also found availability of feedstock is important in determining plant location. Chapter 5 also finds that in the United States state producer tax credits and the federal MTBE ban have a positive effect on ethanol investment (Lin and Thome, 2012). In Europe, competition between plants deters local investments and ethanol support policies encourage investments (Lin and Yi, 2012b). In Canada, competition between plants is enough to deter local investments, the availability of feedstock is important in determining plant location, and the effects of policy support for wheat-based plants are significant (Lin and Yi, 2012a).

Chapter 6 - Cellulosic ethanol has been taken as a promising alternative energy source for the nonrenewable fossil fuel. Cellulose is one of the three primary polymers of lignocellulosic biomass cell wall. It is the most abundant biomaterial on earth. Each cellulose molecule is a linear polymer of thousands of glucose residues. Cellulose can be hydrolytically broken down into glucose either enzymatically by cellulolytic enzymes or chemically by sulfuric or other acids. The produced glucose can be fermented into ethanol. However, structural and compositional features of lignocellulosic biomass such as lignin/hemicellulose shield and crystallinity of cellulose present challenges to efficient hydrolysis of cellulose. Employment of enzymes for the hydrolysis of the lignocellulose is considered the prospectively most viable strategy to offer advantages over other chemical conversion routes of higher yields, minimal byproduct formation, low energy requirements, mild operating conditions, and environmentally friendly processing. It has long-term potential for cost reductions compared to other more established routes such as concentrated acid and two-stage dilute acid hydrolysis, even though the enzymatic route has the highest current costs. Considerable research efforts have been made to improve the bioconversion efficiency of cellulose into ethanol, mainly focusing on three major steps of biomass pretreatment, enzymatic hydrolysis, and fermentation. This chaper reviews the development of biochemical conversion of biomass materials into cellulosic ethanol, including previous achievements, current status, and future work.

Chapter 7 - Alcohol is the most commonly used and abused recreational drug, and one effect of ethanol administration, regardless of whether it is acute or chronic, is the disruption of learning and memory. Although recent studies have demonstrated that ethanol does not produce global deficits but, rather, acts on specific substrates to alter neural function, the understanding of the effects of ethanol on cognitive processes remains incomplete. The studies discussed herein offer support for the specificity of the effects of ethanol on learning-related processes and examine how these effects vary with both the task and the phase of ethanol administration examined. Acute ethanol impairs emotional learning as measured by standard contextual and cued fear conditioning, as well as trace fear conditioning and passive avoidance.

However, the effects of acute ethanol on these tasks are influenced by multiple factors, such as genetics and age. Furthermore, as ethanol administration transitions into chronic and withdrawal from chronic ethanol, the pattern of impairments in emotional learning changes.

This suggests that acute, chronic, and withdrawal from chronic ethanol differentially alter behavior and therefore may also differentially alter neuronal function. Thus, the current review compares and contrasts the effects of acute, chronic, and withdrawal from chronic ethanol within fear conditioning and passive avoidance tasks, and across these two models of aversive/emotional learning.

In: Ethanol: Production, Cellular Mechanisms and Health Impact ISBN: 978-1-62257-297-7
Editors: Roy I. Henry and Billy P. Woods

Chapter 1

MICROORGANISMS AND RAW MATERIALS USED FOR THE PRODUCTION OF BIOETHANOL: MAIN REQUIREMENTS, CURRENT STATUS AND PROSPECTS

Juan C. Higuita, Javier M. Naranjo and Carlos A. Cardona
Universidad Nacional de Colombia sede Manizales
Instituto de Biotecnología y Agroindustria, Manizales, Colombia

ABSTRACT

The worldwide bioethanol demand is rapidly growing and the traditional sources are no longer available in large quantities due to their primary value and use as food. For this reason, other raw materials must be considered and improved microorganisms that increase the fuel ethanol productivity are required. The research associated with the implementation of new raw materials and the search for new fermentative bio-agents is constant. The profitability of the bioethanol production process is highly related to the type of feedstock involved and the used microorganism. One of the main results obtained from research on the bioethanol production is a major knowledge of the fermentation process. This feature has led to many technological advances increasing the potential for processes under no conventional conditions. Nevertheless, these advances have still great limitations and low yielding that impede their scaling up to industrial levels.

The selection of the proper microorganism and feedstock for ethanol fermentation processes should comply not only with the characteristics that maximize the productivity of bioethanol, but also with other important constraints such as no involvement in food security, reduction of undesired by- products and pollutant compounds. The large scale production of bioethanol currently makes use of various microorganisms and raw materials, but a large amount of this biofuel is produced by the fermentation of starch or sugar compounds. New developments include several types of lignocellulosic compounds, such as agroindustrial wastes, grasslands and domestic wastes. The traditional fermenting microorganisms are not very efficient degrading these compounds thus causing the implementation of different pretreatment techniques. Also, no conventional microorganisms capable of utilizing different substrates, like those isolated from natural environments, adapted to different conditions or genetically modified, must be considered.

This chapter is a review of the main raw materials and microorganisms used for the production of bioethanol at industrial and pilot scales, the metabolic routes and cellular mechanisms that govern the conversion of substrates, the requirements that must be met for a cost effective fermentation and the main perspectives related to the bioethanol production process.

INTRODUCTION

Petroleum and its derivatives represent almost 35% of the total primary energy supply in the world and about 60% is employed in the transport sector (*1*). The increase of oil prices, the limited reserves of fossil fuels, the political instability in oil producer countries and the climate change have influenced the development on liquid biofuels. Ethanol produced from renewable sources is considered one of these alternatives, and its production capacity has been undergoing rapid expansion since the beginning of this new millennium. Table 1 shows the main ethanol producer countries and illustrates the increasing trend of fuel ethanol production capacity in the world since 2001 to 2009. Currently, The United States has become the first world producer of ethanol with the 47% of the total ethanol produced in the world in 2009 (see Figure 1).

Many countries have implemented or are implementing programmes for addition of ethanol to gasoline in the transport sector as a renewable oxygenate, to reduce to the oil dependence and improve the combustion processes. For instance, The European Union has issued different directives about the addition of renewable oxygenates to fuels. The oxygenation target of fuels considered the addition of 2 wt.% by 2005 and 5.75 wt.% by 2010. However, the implementation of these directives varies among the different countries. Spain and France are leading the production of bioethanol in Europe. In contrast, Germany has developed the production of biodiesel from rapeseed. In this country, it is considered that the production of fuel ethanol is not economically feasible in comparison to gasoline due to the high costs of feedstocks (e.g., grains, sugar beet, etc) (*3*).

The conventional process of ethanol production can be divided in three main stages: pretreatment of feedstock, fermentation, and separation. The pretreatment of feedstock depends on the type of raw material used; the fermentation conditions are determined by the requirement of microorganisms, operation regime (e.g., batch, fed batch, continuous) and the technological configuration used. In the conventional separation, the ethanol contented in the fermented broth (approximately 10% w/w) can be concentrated until its azeotropic concentration (95.6% w/w) by conventional distillation (*4*). When ethanol is used as carburant mixed with gasoline, it must be in anhydrous form (99,6%) because residual water ethanol can cause engine damage (*5*).

According to the ethanol production cost, the raw material used can represent up to 60% of the total production cost (*6*) since the main raw materials currently used are sugary and starchy materials. The energy consumption cost could account for up to 30% of the total production cost. 80% of the energy consumption is from the downstream processes after fermentation, mainly in the distillation of dilute broth as well as in the treatment of large amount of waste distillage by the multiple evaporation technology (*6*). The profitability of

Table 1. World ethanol production 2001 – 2009 (million liters)

Year	2001	2002	2003	2004	2005	2006	2007	2008	2009
Unites States	7942	9416	11885	14158	16044	19689	25932	36388	42147
Brazil	11525	12735	14560	15344	16115	17714	22243	27095	25323
China	2852	3169	3676	3800	5000	7400	7000	6900	7300
UE	2638	2569	2516	2501	2823	3458	2716	4440	5572
India	1422	1397	1392	1133	1215	1780	2077	2063	1588
Canada	333	330	330	330	355	365	720	950	1370
Other Europe	1262	1322	1379	1335	1379	1288	1334	1283	1257
Other Asia	1102	1010	1078	915	875	809	911	907	932
Other South America	428	411	444	472	454	498	553	615	682
Thailand	77	202	207	247	309	390	411	574	662
Others North & Central America	409	390	407	423	446	519	529	703	473
Others Africa	164	161	182	198	204	220	274	322	393
South African	346	353	358	385	404	409	378	367	385
Oceania	176	182	164	149	149	172	203	221	385
Colombia	1	1	1	1	29	269	275	260	327
Pakistan	42	63	80	100	137	230	350	405	192
Total	30719	33711	38659	41491	45938	55210	66906	83493	88988

Adapted from ASOCAÑA (*2*).

bioethanol production depends on the availability and cost of the raw materials used, the efficiency of the biological conversion process, the type of microorganism used and the configuration of the phase separation of products and byproducts. Although these bioprocesses can be improved in each of the afore mentioned phases, the success of the modifications to the pretreatment steps of raw materials and purification of products depends greatly on the fermentation stage. More specifically, of the metabolic and physiological characteristics of the microorganism used and its ability to transform the substrate to ethanol in an efficient and economic form.

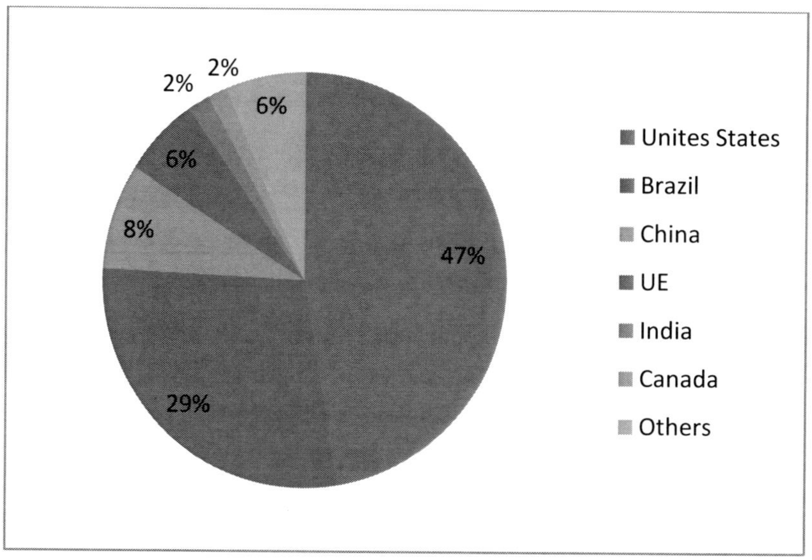

Figure 1. Main Ethanol producer countries in 2009.

To reduce production costs and increase economic and technical feasibility of the production process of ethanol via fermentation, it is important to study the characteristics of microorganisms and available raw materials for ethanol production. In the first part of this chapter, the main raw materials used for ethanol production are analyzed. Then, the microorganisms used for the production of bioethanol at industrial and pilot scales, the metabolic routes and cellular mechanisms that govern the conversion of substrates, are discussed. Finally, an analysis about the improvement of ethanol production is made and the main perspectives related to the bioethanol production process are considered.

RAW MATERIALS

The raw material is one of the most important aspects that determine both the technical and economic viability of the bioethanol producing processes. Bioethanol can be produced from raw materials that contain fermentable simple sugars (such as: sucrose, maltose, glucose, and fructose), glycerol, polysaccharides (starch, cellulose, hemicelluloses) that may be hydrolyzed to producing fermentable sugars. The main raw materials used for ethanol production can be classified according to two different aspects: i) generation of the raw material and ii) the characteristics of their components (molecules).

1) Raw Material Generation

According to their generation, raw materials can be classified into three different categories and Figure 2 explains this general classification.

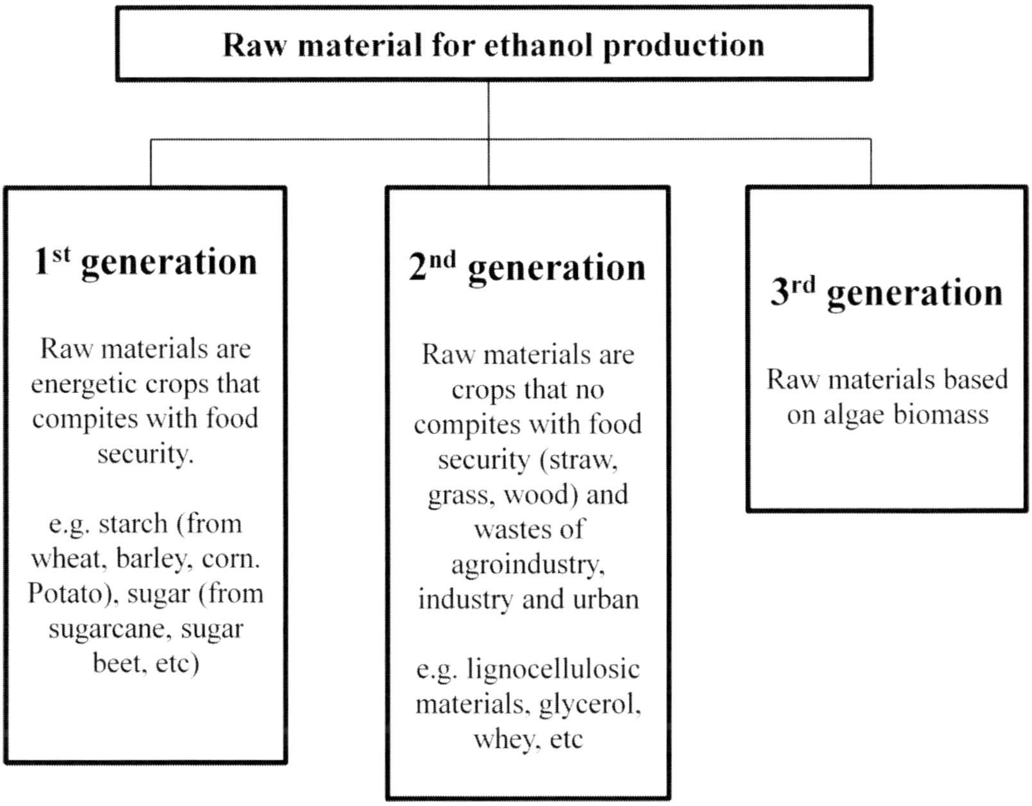

Figure 2. General classification of raw materials for ethanol production.

First Generation

The raw materials used directly from energetic crops such as sugar cane, sugar beet, corn, soy bean, are considered as first generation raw materials. These feedstocks compete directly with food security because they are food crops and/or its production occupies lands that may be used for food production. When producing ethanol from maize (made up from starch chains) or sugarcane (in the form of either cane juice or molasses) the raw material constitutes about 40–70% of the production cost (7).

Second Generation

The main characteristic of this type of raw materials is that they don't compete with food security. They can be crops as straw, grass, wood and agro-industrial, industrial or domestic wastes. The advantage of these feedstocks for ethanol production is their low prices, but the disadvantages are the pretreatment processes (as shredding, densifying, pulverizing, handling and hydrolysis) and transportation. Also, market accessibility and acceptance are hurdles that need to be addressed. Despite these challenges, second-generation feedstocks can produce a much higher amount of fuel ethanol, with the potential for greater savings on greenhouse gas emissions compared to first-generation biofuels. Lignocellulose (a complex polymer composed by three carbohydrates: cellulose, hemicelluloses and lignin) is considered as an attractive second generation feedstock for the production of fuel ethanol, because of its availability in large quantities at low cost. Today the production cost of ethanol from lignocellulose is still high (compared to first generation raw materials), which is the major reason why ethanol from this feedstock has not yet made its breakthrough.

Other important raw material of second generation is glycerol. It is a co-product obtained in the biodiesel production in a weight ratio of 1/10 (glycerol/ biodiesel). The recent growth on the biodiesel industry has carried out a glycerol surplus and its consequent price decrease, which affects directly the biodiesel economy. Producers and refining companies like Dow Chemical, Procter and Gamble Chemicals shut down their glycerol producing plants due to the low glycerol market prices(*8*). Pure glycerol is an important industrial feedstock used in food, drug, cosmetics, pharmaceutical, pulp and paper, leather, textile and tobacco industries (*8*). On the other hand, crude glycerol is refined by larger scale biodiesel producers employing a traditional separation process used to remove impurities. Some of these refining processes are filtration, chemical additions, and fractional vacuum distillation. These processes are expensive and economically unfeasible for small and medium scale plants. Thus, the development of biorefineries based on crude glycerol to produce higher value compounds is necessary in the biodiesel industry to overcome the economic glycerol drawback.

Third Generation

The third generation raw material is derived from microalgae. They are considered to be a viable alternative energy resource that is devoid of the major drawbacks associated with first and second generation biofuels. Microalgae are used specially for biodiesel production due to their capability to render 15–300 times more oil than traditional crops on an area basis (*9*). The resulting cake from microalgal growth is rich in carbohydrates that may be fermented for ethanol production. Although microalgae have a similar biomolecular composition than the traditional crops, they are considered as third generation raw materials since microalgae have a very short harvesting cycle (1– 10 days approximately depending on the process), allowing multiple or continuous harvests with significantly increased yields, compared with conventional crop plants which are usually harvested once or twice a year (*9*).

2) Type of Raw Materials for Ethanol Production According to the Characteristics of their Components

Sugar-Rich Raw Materials

Sugar cane and sugar beets are the main raw materials rich in assimilable sugars for ethanol fermentation. The juice extracted during processing and the obtained molasses (a byproduct of the generation of sugar) are the substrates used by the industries involved in the generation of bioethanol. The cultivation of sugarcane, prevalent in tropical countries like Brazil and Colombia, is the main raw material used for the production of this biofuel. European countries mainly use sugar beet as feedstock.

The integration of sugar refineries with ethanol producer plants is one of the major advantages of ethanol production from molasses, cane juice or sugar beet. This strategy, together with a mass and energy integration, increases the profitability in the production of bioethanol (10).

Some of the sugar-rich raw materials that are used in the production of bioethanol are presented below.

Sugar Cane and Sugar Beet Juice

Cane juice is a greenish liquid obtained during the milling of sugar cane, which is the first stage of sugar production. Its composition varies according to the variety of cane used, the region from which it comes and the grinding conditions. Table 2 shows average values of the composition of sugar cane and sugar beet juice.

Although the content of sugars directly assimilated by the microorganisms is low, the hydrolysis of sucrose is a simple and efficient procedure, which is performed at temperatures of 90°C and pH values of 4, approximately. This is an important requirement to ensure high conversion of sugar to ethanol during the fermentative metabolism of yeasts or other microorganisms that assimilate glucose and/or fructose.

Table 2. Composition of Juice from sugar cane and sugar beet

Component	Concentration (%w/w)
Sugarcane juice	
Water	81-85
Soluble solids	15-19
Salts	0.23-0.67
Proteins	0.08-0.11
Starch	0.05-0.11
Sucrose	12.5-16.72
Glucose	0.3-0.76
Fructose	0.3-0.76
Organic mater (no sugar)	0.5-1
Inorganic compounds	0.2-0.6
Nitrogen compounds	0.5-1
Sugar beet juice	
Solids	17.3
Sucrose	16.5
Raffinose	0.07
Monosaccharide	0.15
Polysaccharides	0.019
Total nitrogen	0.105
Sodium	0.015
Potassium	0.125
Chlorides	0.003
Nitrites	0.005
Nitrates	0.015
Phosphates	0.047
Sulphates	0.0199

Modified from (*11-13*).

Molasses* from *Sugar Cane* and *Sugar Beet

Molasses are a byproduct of the sugar obtaining process. They are, generated during the stages of evaporative crystallization and centrifugation of the mass from the evaporators. It is viscous, thick fudge and its composition of impurities and mineral salts is higher in comparison to raw juices. The composition of this raw material is very variable, it depends on the characteristics of the sugar cane or sugar beet cultivation, the type and efficiency of processing at the mill and the storage characteristics. This implies a great variation in the content of sugars and other nutrients, as well as color, viscosity, flavor and aroma. The composition of molasses from sugar beet and sugar cane is presented in Table 3.

The high osmolality of the fermenting media composed by molasses is a negative feature for ethanol production via fermentation. This osmolality is related to the concentration of sugars and salts in the medium. The principal approach for avoiding the negative influence of salts and other compounds on the fermentation is through the conditioning of molasses by the addition of different compounds that neutralize the inhibitory effects of the medium components. In addition, molasses should be supplemented with nutritional compounds that promote yeast growth. Agents used for cane molasses have demonstrated their usefulness for conditioning beet molasses such as EDTA, ferrocyanide and zeolites (*14*).

Table 3. Composition of cane molasses from sugar beet in %w/w. The compositions are based in a molasses concentration of 79.5°Brix for both cases

Item	Sugarcane	Sugar beet
Specific gravity	1.41	1.41
Total solids	75	77
Total sugars	46	48
Crude protein	3	6
Ash	8.1	8.7
Calcium	0.8	0.2
Phosphorus	0.08	0.03
Potassium	2.4	4.7
Sodium	0.2	1
Chlorine	1.4	0.9
Suphur	0.5	0.5

Modified from (*12*).

According to Castellar et al. (*15*), zeolites could also act as a pH regulator. In the case of fermentations with high glucose concentration and using *Saccharomyces bayanus* as the yeast strain, the zeolite use allows the maintenance of the medium pH around 3.7 leading to the consumption of all the initial glucose and, therefore, to a higher ethanol concentration. Other substances have demonstrated their usefulness during the preparation of media based on molasses from sugar cane are commercial enzymatic complex of amylases, cellulases and amylopectinases allows the conversion of non-fermentable substances into assimilable compounds improving the alcoholic fermentation (*16*). On the other hand, adding a minimum inhibitory concentration of hop acids to molasses will prevent bacterial growth, increase ethanol yields and avoid the need for antibiotics as described in a previously reported patent (*17*).

Starchy Materials

Starch is a polymer exclusively composed of glucose units, which constitutes an important raw material for the production of ethanol. Traditionally, this material has been used for the production of fermented alcoholic beverages, but for its conversion to ethanol it requires the degradation of starch into glucose (12). This type of raw material is extensively used in North America and Europe that use corn, wheat and other grains for ethanol production via fermentation. In tropical countries, other starchy crops (such as tubers) could be implemented on an industrial scale for the generation of alcohol(12). Starch hydrolysis is required to produce ethanol by fermentation. Starch was traditionally hydrolyzed by acids, but the specificity of the enzymes, their inherent mild reaction conditions and the absence of secondary reactions have made the amylases to be the catalysts generally used for this process. α- amylase obtained from thermoresistant bacteria like *Bacillus licheniformis* or from engineered strains of *Escherichia coli* or *Bacillus subtilis* is used during the first step of hydrolysis of starch suspensions. For the amylases to attack starch, these suspensions should be brought to high temperatures (90–110 °C) in order to breakdown the starch kernels. In last years, the possibility of hydrolyzing starch at low temperatures for achieving energy savings is being investigated (*18*). The product of this first step, called liquefaction, is a starch solution containing dextrines and small amounts of glucose. The liquefied starch is subject to saccharification at lower temperatures (60–70 °C) through glucoamylases obtained generally from *Aspergillus niger* or *Rhizopus sp* (*19, 20*).

Corn

Ethanol is produced almost exclusively from corn in the USA. Corn is milled to extract the starch, which is then enzymatically treated to obtain a glucose syrup which is fermented into ethanol. There are two types of corn milling in the industry: wet and dry. During the wet-milling process, corn grain is separated into its components. Starch is converted into ethanol and the remaining components are sold as co-products. During dry-milling, grains are not fractionated and all their nutrients enter the process and are concentrated into a distillation co-product utilized for animal feed called Dried Distiller's Grains with Solubles (DDGS). In general, the liquefaction, saccharification and fermentation steps are the same for both technologies. Fermentation is performed using *Saccharomyces cerevisiae* and is carried out at 30–32 °C with the addition of ammonium sulfate or urea as nitrogen sources. Proteases can be added to the mash to provide an additional nitrogen source for the yeast resulting from the hydrolysis of corn proteins.

Quintero et. al. (*11*) demonstrated that the total operating costs are significantly higher for ethanol production from corn than from sugarcane. In comparison to the corn case, the greater cane demand to produce the same amount of ethanol (about six times higher than the grain requirements) is compensated by the lower cost of this raw material. In fact, the main part of the fuel ethanol costs corresponds to the feedstock: 66.45% and 70.84% using sugarcane and corn, respectively. Usually, the feedstock cost for Brazilian cane ethanol is about 60% of the production costs, whereas for corn (mostly transgenic) is about 63% in USA.

Wheat

Although in France ethanol is mostly produced from beet molasses, it is also produced from wheat by a process similar to that of corn. Some efforts have been made for optimizing fermentation conditions. For example, Wang et al. (*21*) have determined the optimal fermentation temperature and specific gravity of the wheat mash. Soni et al. (*22*) have optimized the conditions for starch hydrolysis using α-amylase and glucoamylase obtained by a solid-state fermentation of wheat bran. To enhance the fermentation performance, high gravity fermentations have been proposed, particularly for the case of wheat mashes. In this kind of process, the initial dissolved solids concentration exceeds 200 g/L implying a higher substrate load. Therefore, higher ethanol concentrations are obtained, lower amounts of process water are required, and energy costs are decreased. The drawbacks of this technology include longer fermentation times, and sometimes incomplete fermentations probably caused by product inhibition, high osmotic pressures and inadequate nutrition (*23*). To accelerate high gravity fermentations, the controlled addition of small amounts of acetaldehyde during the fermentation has allowed the reduction in the cultivation time from 790 h to 585 h without affecting the ethanol yield. It is believed that this positive effect may be caused by the ability of acetaldehyde to replenish the intracellular acetaldehyde pool and restore the cellular redox balance (*23*).

Cassava

Cassava represents an important alternative source of starch not only for ethanol production, but also for production of glucose syrups. In fact, cassava is the tuber that has gained most interest due to its availability in tropical countries being one of the top ten more important tropical crops. Ethanol production from cassava can be accomplished by using either the whole cassava tuber or the starch extracted from it. Starch extraction can be carried out through a high-yield large-volume industrialized process as the Alfa Laval extraction method (FAO, 2004) (*24*), or by a traditional process for small- and mid-scale plants. This process can be considered as the equivalent of the wet-milling process for ethanol production from corn.

The production of cassava with high starch content (85–90% dry matter) and less protein and minerals content is relatively simple. Cassava starch has a lower gelatinization temperature and offers a higher solubility for amylases in comparison to corn starch. The hydrolysis of cassava flour has been proposed for the production of glucose in an enzymatic hollow-fiber reactor with 97.3% conversion (*25*) considering that cassava flour production is more simple and economic than the cassava starch production. However, it is considered that cassava ethanol would have better economic indicators if the whole tuber is used as feedstock, especially when small produccrs arc involved. Fuel ethanol production from whole cassava is equivalent to ethanol production from corn by dry-milling technology. For this, cassava should be transported as soon as possible from cropping areas considering its rapid deterioration due to its high moisture content (about 70%). Hence, this feedstock should be processed within 3–4 days after its harvest. One of the solutions to this problem consists on the use of sun-dried cassava chips (*26*). The farmers send the cassava roots to small chipping factories where they are peeled and chopped into small pieces. The chips are sun-dried during 2–3 days. The final moisture content is about 14% and the starch content reaches 65%. The first step of the process in the distillery is the grinding of the dried cassava chips or fresh roots (if a permanent supply is ensured). Milled cassava is mixed with water and undergoes cooking followed by the liquefaction process. The liquefied slurry is saccharified to obtain glucose, which will be assimilated by the yeast during the next fermentation step. The process can be intensified through a Simultaneous Saccharification Fermentation (SSF) as in the corn case. If fresh roots are employed, a fibrous material is obtained in the stillage after distillation. This material can be used as an animal feed similarly to the DDGS produced in the corn-based process.

Other Starchy Materials

Besides corn and wheat, ethanol can be produced from rye, barley, triticale (*27*), and sorghum (*28*). For these cereals, some pretreatments have proven to be useful. Wang et al. (*27*) have employed the pearling of wheat, barley, rye and triticale grains to increase the starch content of the feedstock in an average of 12% obtaining a 6.5–22.5% enhance in the ethanol yield during fermentation. Ethanol production from bananas and banana wastes using commercial α-amylase and glucoamylase has been studied by Hammond et al.(*29*). In their work, an ethanol yield of 0.5 l EtOH/kg dry matter of ripe bananas was obtained. The processing of starch-containing food wastes by adding malt to the pulverized feedstock has been patented (*30*). One of the most promising crops for fuel ethanol production is the sweet sorghum, which produces grains with high starch contents, stalks with high sucrose and leaves and bagasse with high lignocellulosic contents. In addition, this crop can be cultivated in both temperate and tropical countries requiring only 1/3 of the water needed for cane cropping and half of the water required by corn. Moreover, it is tolerant to the drought, flooding and saline alcalinity(*31*). Grassi (*32*) reports that from some varieties of sweet sorghum, the following productivities can be obtained: 5 ton/Ha grains, 8 ton/Ha sugar and 17 ton dry matter/Ha lignocellulosics. The estimated price for fuel ethanol production from this feedstock is US$200–300/ m^3, whereas the corresponding one for sugar cane ethanol is US$260, for corn ethanol is US$300–420 and for lignocellulosic ethanol is US$450.

Lignocellulosic Biomass

The lignocellulosic complex is the most abundant biopolymer on earth. It is considered that lignocellulosic biomass comprises about 50% of the world biomass and its annual production was estimated in 10–50 billion ton (*33*). In general, prospective lignocellulosic materials for fuel ethanol production can be divided into six main groups as follows: crop residues (cane bagasse, corn stover, wheat straw, rice straw, rice husk, barley straw, sweet sorghum bagasse, olive stones and pulp), hardwood (aspen, poplar), softwood (pine, spruce), cellulose wastes (newsprint, waste office paper, recycled paper sludge), herbaceous biomass (alfalfa hay, switchgrass, reed canary grass, coastal Bermudagrass, thimothy grass), and municipal solid wastes (MSW). The composition of some of these materials are shown in Table 4. (e.g. Sun and Cheng (*34*)). Numerous studies for developing large-scale production of ethanol from lignocellulosics have been carried out in the world. However, the main limiting factor is the higher degree of complexity inherent to the processing of this feedstock.

Two of the main polymers (cellulose and hemicelluloses) of the biomass should be broken down into fermentable sugars in order to be converted into ethanol or other valuable products. This degradation process is complicated, energy-consuming and still non-completely developed.

Table 4. Composition of some lignocellulosic materials used for ethanol production

Group of Lignocellulosic materials	Examples	Cellulose (%)	Hemicellulose (%)	Lignin (%)
Crop residues	Corn cobs	45	35	15
	Nut shells	25-30	25-30	30-40
	Leaves	15-20	80-85	0
	Wheat-straw	30	50	15
	Empty fruit bunches[1]	14	13	8
	Rice husk[1]	30,87	32,13	35
	Coffee cut stems[1]	33,06	24,94	22
	Sugar cane bagasse[1]	30,8	25,2	21
Hard wood	Hard wood stems	40-55	24-40	18-25
Softwood	Softwood stems	45-50	25-35	25-35
Herbaceous biomass	Grasses	25-40	35-50	10-30
	Coastal Bermuda grass	25	35.7	6.4
	Switch grass	45	31.4	12
Cellulose wastes	Newspaper	40-45	25-40	1830
	Waste papers from chemical pulps	60-70	10-20	5-10
	Primary wastewater solids	8-15	NA[2]	24-29
	Paper	85-99	0	0-15

1: Based on results obtained by Quintero (*3*), 2: NA: No Available
Adapted from Sun and Cheng (*34*).

Lignocellulosic material consists of mainly three different types of polymers, namely cellulose, hemicellulose and lignin, which are associated which each other. Cellulose is composed by D-glucose subunits, linked by b-1,4 glycosidic bonds (Figure 3). The cellulose in a plant consists of parts with a crystalline (organized) structure, and parts with a, not well-organized, amorphous structure. The cellulose strains are 'bundled' together and form the so called cellulose fibrils or cellulose bundles. These cellulose fibrils are mostly independent and weakly bound through hydrogen bonding (*35*).

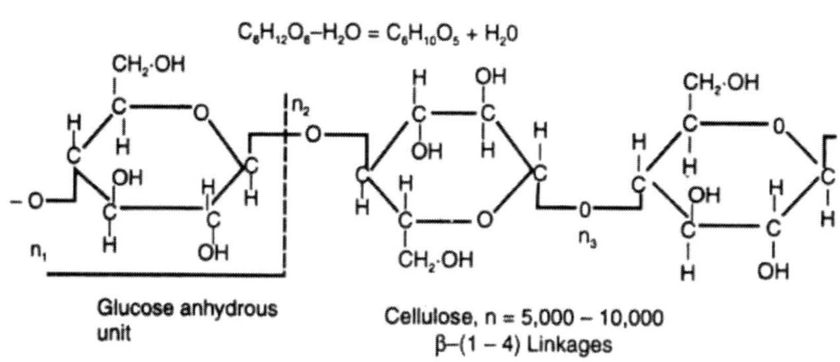

Figure 3. A generalized structure of cellulose.

Hemicellulose is a complex carbohydrate structure that consists of different polymers like pentoses (i.e., xylose and arabinose), hexoses (like mannose, glucose and galactose), and sugar acids (Figure 4). The dominant component of hemicellulose from hardwood and agricultural plants, like grasses and straw, is xylan, while this is glucomannan for softwood (*36*). Hemicellulose has a lower molecular weight than cellulose and branches with short lateral chains that consist of different sugars, which are easy hydrolyzable polymers. Hemicellulose serves as a connection between the lignin and the cellulose fibers and gives the whole cellulose–hemicellulose– lignin network more rigidity. Considering cellulose, hemicelluloses and lignin, the hemicellulose is the most thermal-chemically sensitive component. During thermal–chemical pretreatment the side groups of hemicellulose react first, followed by the hemicellulose backbone.

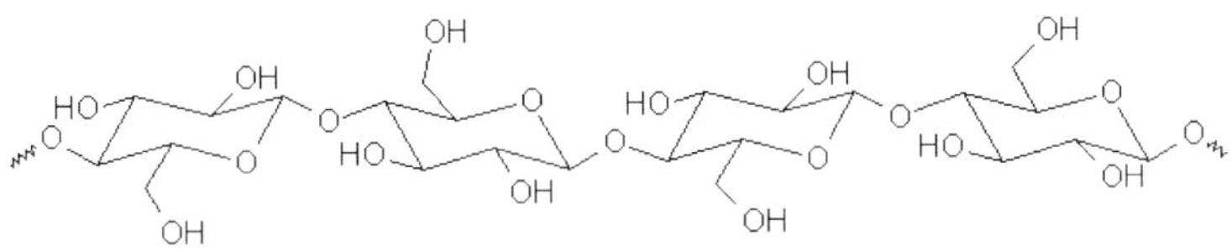

Figure 4. A generalized structure of hemicelluloses.

Lignin is present in the cellular wall and is, after cellulose and hemicellulose, one of the most abundant polymers in nature. It is an amorphous heteropolymer consisting of three different phenylpropane units (see Figure 5) that are held together by different kinds of linkages. The main purpose of lignin is to give the plant structural support, impermeability and resistance against microbial attack and oxidative stress. The amorphous heteropolymer is also non-water soluble and optically inactive; all this makes the degradation of lignin very tough. Lignin, just like hemicellulose, normally starts to dissolve into water around 180 °C under neutral conditions (*37*). The solubility of the lignin in acid, neutral or alkaline environments depends, however, on the precursor (p-coumaryl, coniferyl, sinapyl alcohol or combinations of them) of the lignin.

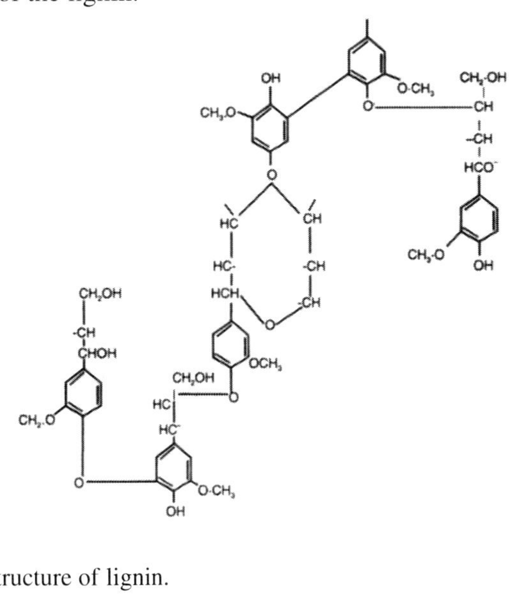

Figure 5. A generalized structure of lignin.

Pretreatment of Lignocellulosic Biomass

The main processing challenge in the ethanol production from lignocellulosic biomass is the feedstock pretreatment. The lignocellulosic complex is made of a matrix of cellulose and lignin bound by hemicellulose chains. During the pretreatment, this matrix should be broken in order to reduce the crystallinity degree of the cellulose and increase the fraction of amorphous cellulose, the most suitable form for enzymatic attack (Figure 6). Additionally, the main part of the hemicellulose should be hydrolyzed and lignin should be released or even degraded. The fact that the cellulose hydrolysis is affected by the porosity (accessible surface area) of lignocellulosic materials should be also considered.

The yield of cellulose hydrolysis is less than 20% of the theoretical value when pretreatment is not carried out, whereas the yield after pretreatment often exceeds 90% of this value(*38*). Therefore, the aim of the pretreatment is the removal of lignin and hemicellulose, the reduction of crystalline cellulose and the increase in the porosity of the materials. Additionally, the pretreatment should improve the formation of sugars or the ability to form them during the succeeding enzymatic hydrolysis, and avoid the formation of inhibitors for subsequent hydrolysis and fermentation processes. For the pretreatment of lignocellulosics, several physical, physical–chemical, chemical and biological processes have been proposed and developed (*34*).

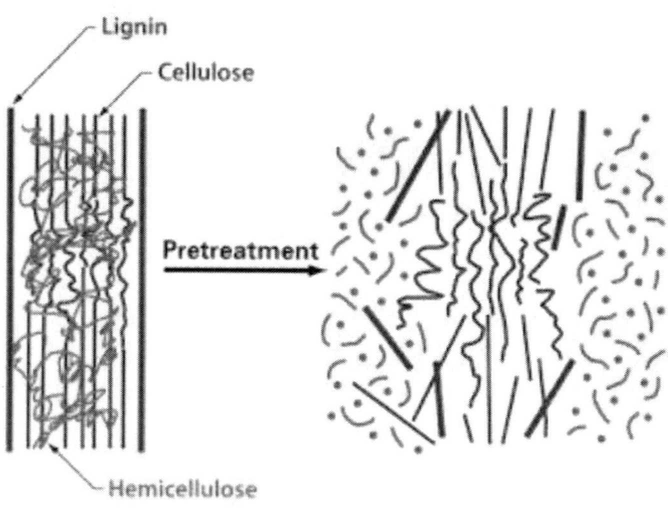

Figure 6. Pretratment effect in the lignocellulosic structure.

Physical Methods

Waste materials can be comminuted by a combination of chipping, grinding and milling to reduce cellulose crystallinity. This reduction facilitates the access of cellulases to the biomass surface increasing the conversion of cellulose. The energy requirements of mechanical comminution of lignocellulosic materials depend on the final particle size and biomass characteristics. Although mechanical pretreatment methods increase cellulose reactivity towards enzymatic hydrolysis, they are unattractive due to their high energy and capital costs (*39*). Pyrolysis has also been tested as a physical method for pretreatment of lignocellulosic biomass since cellulose rapidly decomposes when is treated at high temperatures.

Physical–Chemical Methods

Physical–chemical pretreatments are considerably more effective than physical methods. The steam explosion is the most studied method of this sort. During this process, the use of saturated steam at high pressure causes autohydrolysis reactions in which part of the hemicellulose and lignin are converted into soluble oligomers. The factors affecting steam explosion pretreatment are residence time, temperature, chip size and moisture content. To consider the combined action of both temperature and time over the performance of steam explosion pretreatment, the so-called severity index has been defined including a correction term when this process is carried out under acidic conditions (*40, 41*). In some cases (e.g. herbaceous waste), the use of very small particles is not desirable considering the economy of the process (*42*).This method is recognized as one of the most cost-effective for hardwood (e.g., poplar, oak, birch, maple) and agricultural residues, but is less efficient for softwood (e.g., pine, cedar). Shahbazi et al. (*40*) proposed a fractionation procedure for softwood based on steam explosion and alkaline delignification in order to produce ethanol and related co-products.

One of the most promising methods is the pretreatment with Liquid Hot Water (LHW) or thermohydrolysis. Laser et al. (*43*) mention that under optimal conditions, this method is comparable to dilute acid pretreatment but without addition of acids or production of neutralization wastes. In addition, this technology presents elevated recovery rates of pentoses and does not generate inhibitors (*44*). Nevertheless, solid load is much less than for steam explosion, which is usually greater than 50%. Another physical–chemical method is the Ammonia Fiber Explosion (AFEX) process whose fundament is similar to steam explosion. As AFEX method and steam explosion, CO_2 explosion uses the same principle but the yields are relatively low (*34*).

Chemical Methods

Chemical pretreatments employ different chemical agents as ozone, acids, alkalis, peroxide and organic solvents. Inorganic acids as H_2SO_4 and HCl have been preferably used for biomass pretreatment. Hydrolysis with dilute sulfuric acid has been successfully developed given that high reaction rates can be achieved improving significantly the subsequent process of cellulose hydrolysis. In contrast, the costs of dilute acid pretreatment are higher than the corresponding ones of steam explosion or AFEX process (*34*). The National Renewable Energy Laboratory (NREL) of the US Department of Energy, which currently is developing ethanol production technologies from biomass, has preferred the dilute acid pretreatment for the design of its process alternatives (*45, 46*). Lynd (*38*) points out that the main advantage of this process related to steam explosion is the higher recovery of sugars derived from hemicellulose. For hardwood, this recovery is about 80% for dilute acid pretreatment, and does not exceed 65% for steam explosion. Dilute acid pretreatment can also be accomplished in a two-stage way. For this, a first depolymerization stage of hemicellulose at 140ºC during 15 min is carried out in order to avoid the formation of furan compounds and carboxylic acids, followed by a second stage at 190ºC during 10 min to make cellulose more accessible to enzymatic hydrolysis. Some authors (*47, 48*) point out that the realization of dilute-acid pretreatment at low temperatures (121ºC) allows avoiding the degradation of sugars to furfural and hydroxymethylfurfural (HMF), but the sugars yields are lower.

Concentrated acids also have been used for pretreatment. A fuel ethanol production process from cane bagasse involving the pretreatment with concentrated sulfuric acid has been patented (*49*). This technology implies the retrofitting of sugar mills in order to produce ethanol and improve energetic indexes of this kind of processes (*50*).

Alkaline pretreatment is based on the effects of the addition of dilute bases on the biomass such as: increase of internal surface by swelling, decrease of the polymerization degree and crystallinity, destruction of links between lignin and other polymers, and breakdown of lignin. The effectiveness of this method depends on the lignin content of the biomass (*34*). In general, the utilization of bases as sodium hydroxide or solvents such as ethanol or methanol (organosolv process) allows the dissolution of lignin, but their costs are so high that these methods are not competitive for large scale plants (*51*).

Biological Methods

Biological pretreatment has low energy requirements and mild environmental conditions. However, most of these processes are too slow limiting its application at industrial levels. Many white-rot fungi degrade the lignin and, for this reason, they have been utilized for ligninases production and lignocellulose degradation. The fermentation processes for producing the enzymes can be both by submerged culture and solid-state fermentation. The fungus *Phanerochaete chrysosporium* has been proposed in the patent of Zhang (*52*) for degrading the lignin in a biomass-to-ethanol process scheme involving the separate fermentation of pentoses and hexoses. Tengerdy and Szakacs (*53*) and Kang et al. (*54*) highlight the viability of producing cellulases and hemicellulases by solid-state fermentation.

According to preliminary evaluations of the NREL, the cost of cellulases produced *in situ* by submerged culture is US$0.38/100,000 FPU (Filter Paper Units, a way for measuring cellulase activity). Thus, cellulase costs comprise 20% of the ethanol production costs assuming them in US$1.5/gallon. On the other hand, the commercial cellulose cost (US$16/100,000 FPU) is prohibitive for this process. In contrast, these authors indicate that the cost of producing cellulases by solid-state fermentation of corn stover could reach up to US$0.15/100,000 FPU that would correspond to US$0.118/gal EtOH, i.e. near 8% of the total costs.

Hydrolysis of Cellulose

For fermentation of lignocellulosic materials, cellulose should be degraded into glucose (saccharification) using acids or enzymes. In the former case, concentrated or dilute acids can be used. If dilute acids (i.e., H_2SO_4 and HCl) are employed, temperatures of 200–240 °C at 1.5% acid concentrations are required to hydrolyze the crystalline cellulose, but the degradation of glucose into HMF and other non-desired products is unavoidable under these conditions. Similarly, xylose is degraded into furfural and other compounds. During a two-stage regime, a first stage under mild conditions (190 °C, 0.7% acid, 3 min) is carried out to recover pentoses, while in the second stage, the remaining solids undergo harsher conditions (215 °C, 0.4% acid, 3 min) to recover hexoses. In this way, a 50% glucose yield is obtained (*55*). One variant of the acid hydrolysis is the employment of extremely low acid and high temperature conditions during batch processes (auto-hydrolysis approach) that has been applied to sawdust (*56*). A concentrated acid process using 30–70% H_2SO_4 has higher glucose yields (90%) and is relatively rapid (10–12 h) but the amount of used acid is a critical economic factor. By continuous ion-exchange, over 97% acid recovery is possible (*55*).

However, cellulose hydrolysis is currently carried out using microbial cellulolytic enzymes. Enzymatic hydrolysis has demonstrated better results for the subsequent fermentation because no degradation components of glucose are formed although the process is slower. Most of the commercial cellulases are obtained aerobically from *Trichoderma reesei*, though a small portion is obtained from *Aspergillus. niger*. *T. reesei* releases a mixture of cellulases, among which at least two cellobiohydrolases, five endoglucanases, β-glucosidases and hemicellulases can be found (*57*). The action of cellobiohydrolases causes a gradual decrease in the polymerization degree while endoglucanases cause the rupture of cellulose in smaller chains reducing rapidly the polymerization degree.

Endoglucanasesespecially act on amorphous cellulose, whereas cellobiohydrolases can act on crystalline cellulose as well (*58*). Although *T. reesei* produces some β-glucosidases, which are responsible of hydrolyzing the formed cellobiose into two molecules of glucose, their activities are not very high. Unfortunately, cellobiohydrolases are inhibitedby the cellobiose. For this reason, β-glucosidase from other source needs to be added in order to complement the action of the cellulases of this fungus. Factorial optimization techniques have been applied for the design of cellulases mixtures from different sources including β-glucosidase in order to maximize the yield of produced glucose (*59*). The development of a multicellulase plasmid in which the different cellulase genes could be expressed to produce cellulases with an optimum ratio from a single cultivation has been suggested. Cellulases should be adsorbed on the surface of substrate particles before hydrolysis of insoluble cellulose takes place. The three-dimensional structure of these particles in combination with their size and shape determines whether β-glucosidic linkages are or are not accessible to enzymatic attack (*57*). This makes cellulose hydrolysis to be slower compared to the enzymatic degradation of other biopolymers. For instance, the hydrolysis rate of starch by amylases is 100 times faster than the hydrolysis rate of cellulose by cellulases under industrial processing conditions.

Detoxification of Lignocellulosic Hydrolyzates

During pretreatment and hydrolysis of lignocellulosic biomass, a great amount of compounds that can seriously inhibit the subsequent fermentation are formed in addition to fermentable sugars. Inhibitory substances are generated as a result of the hydrolysis of the extractive components, organic and sugar acids esterified to hemicellulose (acetic, formic, glucuronic, galacturonic acids), and solubilized phenolic derivatives. In the same way, inhibitors are produced from the degradation products of soluble sugars (furfural, HMF) and lignin (cinnamaldehyde, p-hydroxybenzaldehyde, syringaldehyde), and as a consequence of corrosion (metal ions) (*38, 60*). For this reason and depending on the type of the employed pretreatment and hydrolysis, a detoxification of the streams that will undergo fermentation is required. Detoxification methods can be physical, chemical or biological. As pointed out by Palmqvist et. al. (*61*), these methods cannot be directly compared because they vary in the neutralization degree of the inhibitors. In addition, the fermenting microorganisms have different tolerances to the inhibitors. In the model processes developed by NREL, ionic exchange (*46*) and overliming (*45*) have been proposed as the detoxification methods. Alkali treatment is considered one of the best detoxification methods. By this method, furaldehydes and phenolic compounds are mainly removed leading to great improvement in fermentability, especially in the case of dilute-acid hydrolyzates (*62*). Treatment with calcium hydroxide (overliming) or ammonia has shown better results than treatment with sodium or potassium hydroxide, but this difference has not been understood.

Other very diverse detoxification methods have been proposed as: neutralization with lime followed by the addition of activated carbon and filtration for acetic acid removal; partial removal of acetic acid, furfural and soluble lignin by molecular sieves; vapor stripping for removal of volatile inhibitors (*63*); and adsorption using activated carbon, diatomite, bentonite and zeolite after neutralization or overliming (*64*).

An alternative biological method for detoxification of dilute solutions resulting from biomass pretreated by pyrolysis has been proposed (*65*). It is based on a biofilm reactor that uses a mixed culture of aerobic bacteria naturally immobilized on a plastic support. In this way, the biofilm-associated cells are more resistant to the toxic substances released during the biomass pretreatment. The use of extraction with supercritical fluids has also been proposed for detoxification of wood hydrolyzates (*66*). The presence of inhibitors directly influences the course of ethanolic fermentation. In continuous or fedbatch fermentations, feed of the bioreactors is carried out with not very high flow rates allowing a low concentration of inhibitors in the broth. In continuous systems, inhibitors diminish the growth rate and, therefore, the process productivity that is directly linked to the dilution rate. In systems with cell retention (e.g., by cell recirculation using filtration, sedimentation or centrifugation), the increase of accumulables, including the inhibitors, makes the productivity to fall down imposing the need of implementing purge streams. Taherzadeh (*67*) developed a simple strategy for on-line feedback control of fed-batch cultivation for *in situ* detoxification of spruce and birch hydrolyzates. Through this strategy, the same yeast cells converted the inhibitors and maintained their concentration at low levels without the need of any detoxification treatment. Thus, the maximal specific productivity of ethanol increased in more than 10 times (*68*). Purwadi et al. (*69*)employed a continuous cultivation system using a flocculating strain of *S. cerevisiae* to ferment a non-detoxified spruce hydrolyzate. The obtained results demonstrated that a high-cell system with recycling of cells allow the *in situ* detoxification of the pretreated biomass at high dilution rates without the need of any detoxification method. It has shown the possibility of converting the hydrolyzates into ethanol in two hours (at dilution rates of 0.5 h_1), which represents an important outcome in the cultivation of toxic pretreated lignocellulosic biomass. Most of the studies on the effect of toxic compounds on growth and ethanol production have been performed for *S. cerevisiae* and xylose-fermenting yeasts. Palmqvist et al. (*70*)carried out extensive experiments to assess the effect of acetic acid, furfural and p-hydroxybenzoic acid on growth and ethanol productivity of *S. cerevisiae* and *Candida shehatae* through a full factorial design. Oliva et al. (*71, 72*) performed the study of the effect of compounds released during the pretreatment of poplar biomass by steam explosion for the thermotolerant yeast *Kluyveromyces marxianus*, showing that growth is more affected than ethanol production, and that this microorganism exhibits an important resistance to aromatic compounds.

Recombinant microorganisms have also been investigated regarding their capacity for fermenting lignocellulosic hydrolyzates containing inhibitors. Zaldivar et al.(*73, 74*) analyzed the effect of several aldehydes and alcohols formed during the dilute-acid pretreatment on the growth and fermentation of a genetically modified *E. coli* capable of assimilating xylose. In the case of aldehydes, these authors showed that only furfural strongly inhibited ethanol production. One new approach to tackle the presence of inhibitors in biomass hydrolyzates is the development of inhibitor-tolerant strains of microorganisms by means of genetic modification and metabolic engineering.However, Belkacemi et al. (*75*) point out that due to the synergistic interactions among inhibitors and the lack of information about the mechanisms of these interactions, it is not clear against what inhibitor resistance is desired. In this way, intense efforts are being carried out for the identification of inhibitory substances, as well as the determination of their inhibition mechanisms. Palmqvist and Hahn-Hägerdal (*60*) have reviewed the main works carried out in this field applied to wood hydrolyzates. These authors emphasize that these studies will allow the minimization of inhibitors formation during pretreatment and hydrolysis, the prediction of hydrolyzates fermentability and the development of more efficient detoxification methods.

One of the main problems during the pretreatment and hydrolysis of biomass is the variability in the content of lignin and hemicellulose. This variability depends on factors as the type of plant from which the biomass is obtained, crop age, method of harvesting, etc. This feature results in that no one of the previously described pretreatment methods could be applied in a generic way for many different feedstocks (*33*). The future trends for improving the pretreatment of lignocellulosic feedstocks also include the production of genetically modified plant materials with higher carbohydrate content or modified plant structure to facilitate pretreatment in milder conditions or using hemicellulases. It is estimated that the use of these new materials along with improved conversion technologies, could reduce the ethanol cost from lignocellulosic biomass in US$0.11/L in the next ten years (*46*).

Glycerol

The glycerol molecule (1,2,3-propanetriol) is a highly reactive tri-alcohol which has two primary and a secondary hydroxyl groups. Glycerol is a water soluble, colorless, odorless, viscous, and hygroscopic liquid with a specific gravity of 1.261 g mL^{-1}, melting temperature of 18.2ºC, and a boiling temperature of 290ºC (accompanied by decomposition). Chemically, glycerol is available for reacting with a stable alcohol under most operation conditions, and it is basically non-toxic to human health and to the environment. The key feature of its usefulness is the particular combination among its physicochemical properties, compatibility with other substances, and easy handling. Due to these particular properties, glycerol has found more than 1500 end-uses or large volume applications. Glycerol is abundant in nature as the structural component of many lipids. It is produced by yeasts during osmoregulation to decrease extracellular water activity due to its compatible solubility (*76*). Wide glycerol occurrence in nature allows different kinds of microorganisms to metabolize it as a sole carbon and energy source. Glycerol can substitute traditional carbohydrates, such as sucrose, glucose, and starch, in some industrial fermentation processes (*77-79*). One of many promising applications to take advantage of the glycerol surplus is its bioconversion to high value compounds through microbial fermentation.

Glycerol is a molecule of high reducing power that can be used as energy source in microorganisms. Some microorganisms like *E. coli* have shown the ability to metabolize it in the presence of external electron acceptors. The glycerol degradation process begins with the GlpF gene incorporation into the cytoplasm. A later phosphorylation process is carried out which is catalyzed by the GlpK kinase. The phosphorylated carbohydrate (i.e., glycerol 3-phosphate) starts an oxide-reduction process which is accelerated by different enzymes.

A comparison between ethanol production in terms of manufacturing requirements, feedstocks, and operational costs, from corn or glycerol (including H_2 and CO_2 or formic acid as co-products) was presented by Yazdani and González (*80*). Figure 3 shows a comparative scheme for ethanol production from corn and glycerol.

The ethanol production scheme is more complex for corn than for glycerol and also a higher capital investment is required. Moreover, as discussed by Yazdani and González (*80*) it is expected that the operational costs are almost 40% lower for ethanol production from glycerol, even when no credit was given to the formate or hydrogen co-produced with ethanol.

ETHANOL PRODUCER MICROORGANISMS

There are many microorganisms reported in literature that can produce ethanol. Fungi, bacteria, and yeast are the most commonly used microorganisms that produce bioethanol via for fermentation.

Yeasts

There are different types of ethanol producer yeasts. Each one of them has the ability to metabolize a particular substrate or group of substrates, and the oxygen requirement varies between species. Table 5 shows the main types of ethanol producer yeasts, their respective substrates and fermentation conditions.

Xylose-fermenting yeasts, such as *Pichia. stipitis*, *Pachysolen tannophilus* and *C. shehatae* require precisely regulated oxygenation for maximal ethanol production (*81, 82*), but their ethanol production rate from glucose is at least five times less than that observed for *Saccharomyces cerevisiae*. Moreover, their culture requires oxygen and the ethanol tolerance is 2-4 times lower (*33*). Increased cultivation temperature accelerates metabolic processes and lowers the refrigeration requirements. Yeast as *Kluyveromyces marxianus* has been tested as a potential ethanol producer at temperatures higher than 40ºC (*83*). Due to the fact that the main sources of xylose are hydrolysates from lignocellulosic materials, these yeasts require a detoxification of the hydrolysate because they withstand the inhibitory environment of lignocellulose hydrolysates poorly (*84, 85*). These factors increase the cost of xylose fermentation.

S. cerevisiae has an efficient anaerobic sugar metabolism, tolerates inhibitory industrial substrates better than other microorganisms (*88, 91*) and ferments hexoses abundantly present in lignocellulosic, starch and sugar hydrolysates, such as glucose, mannose and galactose with high yield and productivity.

Currently, the yeast *S. cerevisiae* is the principal microorganism used for ethanol production from raw materials rich in sugar and starch. The main metabolic pathway involved in the ethanol fermentation by this microorganism is the Embden–Meyerhof–Parnas (EMP) pathway, through which one molecule of glucose is metabolized and two molecules of pyruvate are produced. Figure 7 shows this pathway.

Table 5. Summary of the main yeast strains used for ethanol production

Yeast strain	Substrates	Fermentation conditions	Remarks	Reference
Saccharomyces cerevisiae	Glucose, fructose, sucrose, Maltose, Maltotriose, Galactose, Mannose, Glucose	Anaerobic 30-37°C	Tolerates inhibitory industrial substrates better than other microorganisms	(*86-88*)
Pichia stipitis	Glucose Xylose	Micro-aerobic 26-35°C	Require well regulated oxygenation for maximal ethanol production. Require detoxification of the hydrolysate. Ethanol production rate from glucose is at least five times less than that observed for *S. cerevisiae*	(*81, 82*)
Pachysolen tannophilus	Xylose Glycerol	Micro-aerobic	Require well regulated oxygenation for maximal ethanol production. Require detoxification of the hydrolysate. Ethanol production rate from glucose is at least five times less than that observed for *S. cerevisiae*.	(*81, 82*)

Table 5. (Continued)

Yeast strain	Substrates	Fermentation conditions	Remarks	Reference
Candida shehatae	Glucose Xylose	Micro-aerobic 20-31°C	Require well regulated oxygenation for maximal ethanol production. Require detoxification of the hydrolysate. Ethanol production rate from glucose is at least five times less than that observed for *S. cerevisiae*.	(*81, 82*)
Kluyviromyces marxianus	Glucose	Anaerobic 40-45°C	Uses a directly cellulose for ethanol production. Has high ethanol productivity from cellulose (0.5 g of ethanol/g cellulose in 78 h).	(*89*)
Kluyveromyces fragilis	-	-	Uses a directly cellulose for ethanol production. Has high ethanol productivity from cellulose	(*89*)
Candida acidothermophilum	-	-	Thermotolerant yeast. Produces 80% of the theoretical ethanol yield	(*90*)

Modified from Quintero (*3*).

Table 6. Summary of the main bacterial strains used for ethanol production

Bacterial strain	Substrates	Fermentation conditions	Remarks	Reference
Zymomona mobilis	Fructose, sucrose, glucose	Anaerobic 30°C	Has high growth rate and specific ethanol production. Its fermentable carbohydrates are limited to glucose, fructose and sucrose. Has a high ethanol tolerance	(*96, 97*)
Zymobacter palmae	-	-	Facultative anaerobe bacterium. Use a broad range of carbohydrate substrates (hexoses, α-linked di- and tri-saccharides, and sugar alcohol). Shows productivity similar to that of *Z. mobilis.*	(*97*)
Recombinant E. coli	-	-	Strain capable of assimilating both hexoses and pentoses.	(*98*)
Thermoanaerobacter thermosaccharolyticum	Glucose Xylose	Anaerobic 60°C	These bacteria may transform pentoses and aminoacids into ethanol. Can synthesize up to 2 mol EtOH/mol hexose. Main drawback consists in their very low ethanol tolerance.	(*99*)
Clostridium thermocellum	Cellulose Glycerol	Anaerobic 55-65°C	These bacteria may transform pentoses and aminoacids into ethanol. Can synthesize up to 2 mol EtOH/mol hexose. Can even directly convert lignocellulosic materials into ethanol Main drawback consists in their very low ethanol tolerance	(*99*)

Modified from Quintero (3).

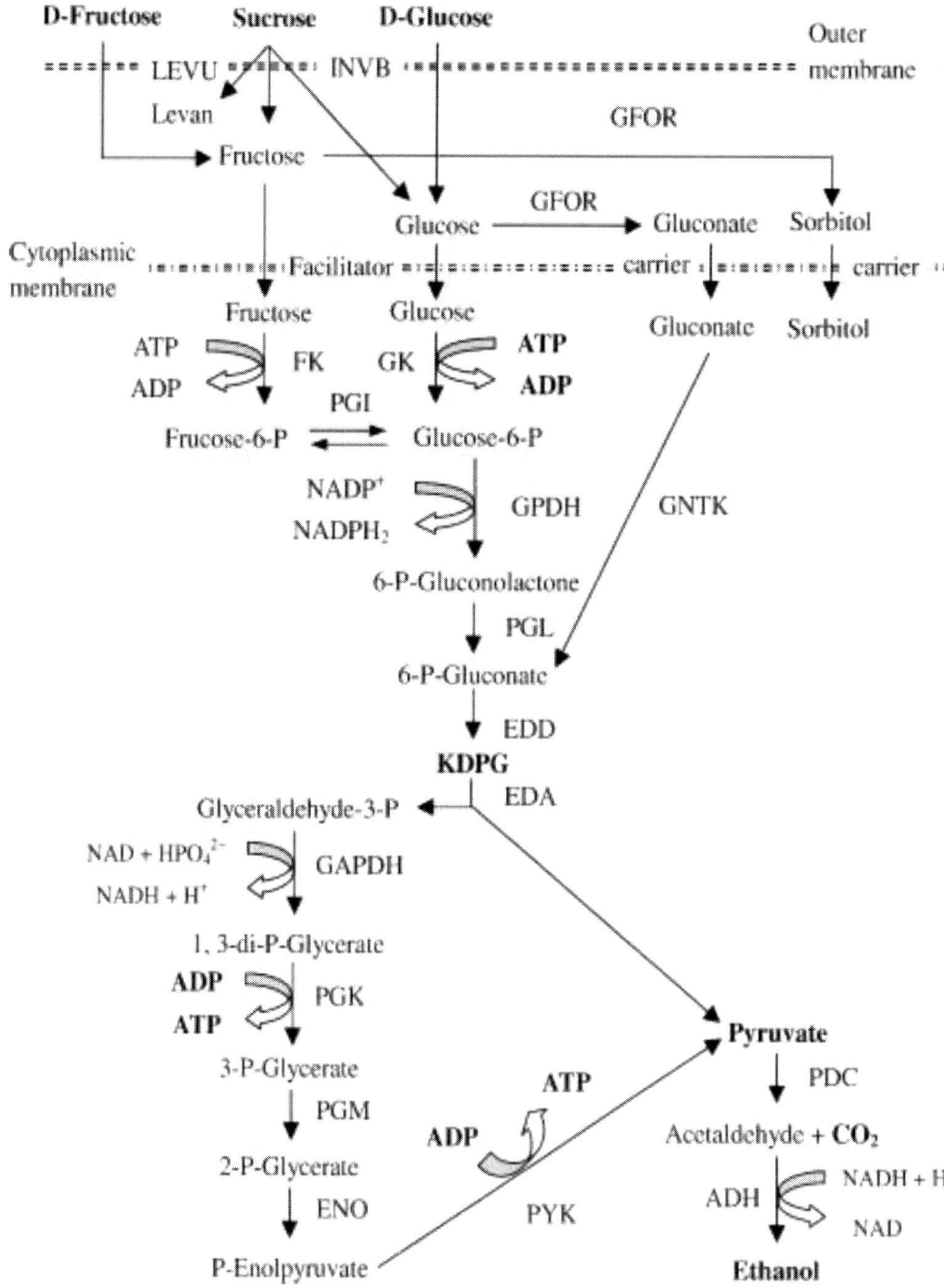

Figure 7. The Embden–Meyerhof–Parnas methabolic pathway, used by *Saccharomyces cerevisiae* in the ethanol production process(*92*).

Equation 1 explains the global reaction by the EMP pathway.

$$C_6H_{12}O_6 + 2P_i + 2ADP \rightarrow 2C_2H_5OH + 2CO_2 + 2H_2O + 2ATP \quad \text{(Eq.1)}$$

Theoretically, 100 g of glucose will produce 51.4 g of ethanol and 48.8 g of carbon dioxide. However, in practice, microorganisms use some of the glucose for growth and the actual yield is less than 100% of the theoretical value, due to the following reasons:

- The 2 ATP generated in glycolysis are used immediately for cell biosynthesis. Thus, growth is strongly linked to ethanol production and the biomass can be considered a by-product in the fermentation of glucose by *S. cerevisiae (92)*.
- Intermediates of the glycolytic pathway are deviated to generate other by-products such as glycerol, organic acids and higher alcohols (*10*).
- Yeast cells suffer from various kinds of stresses during ethanol fermentations. Many of them are synergistic, affecting yeast cells more severely than any single one, leading to reduced yeast viability and vigor as well as lower ethanol yields. The stresses can be from the yeast cell metabolism such as ethanol accumulation and its corresponding inhibition on yeast cell growth and ethanol production. Other stresses can be environmental such as nutrient deficiency, high temperature and contamination.
- Bacterial contamination is known to be a major cause of reduction in the ethanol yield during ethanol production from molasses because of sugar consumption by bacteria. Such bacteria also produce a by-product which inhibits yeast growth. *Lactobacillus* sp. and *Bacillus* sp. may be the most harmful bacteria that contaminate molasses because of their rapid growth abilities, especially *Lactobacillus* sp. because it is tolerant to high temperature and low pH (*93*).

The ethanol production yields observed in the industry can vary between 90-93% of the reported theoretical yield (*92*).

Fermentation of Pentoses

One of the main problems in bioethanol production from lignocellulosics is that *S. cerevisiae* can ferment only certain mono- and disaccharides like glucose, fructose, maltose or sucrose. This microorganism is not able to assimilate cellulose and hemicellulose directly. In addition, pentoses obtained during hemicellulose hydrolysis (mainly xylose) cannot be assimilated by this yeast. A way to overcome this obstacle is through recombinant DNA technology (genetic engineering). Other approach to this problem is the use of pentose fermenting microorganisms like some species of yeasts and bacteria. In this case, configurations involving the separate fermentation of pentoses and hexoses have been proposed. Pentose fermenting yeasts require a careful control for maintaining low oxygen levels in the culture medium needed for their oxidative metabolism. Additionally, these yeasts successfully ferment pure xylose but not the aqueous hemicellulose streams generated during the biomass pretreatment, probably due to the presence of different inhibitors (*94*). Figure 8 shows a metabolic pathway for ethanol production from pentoses.

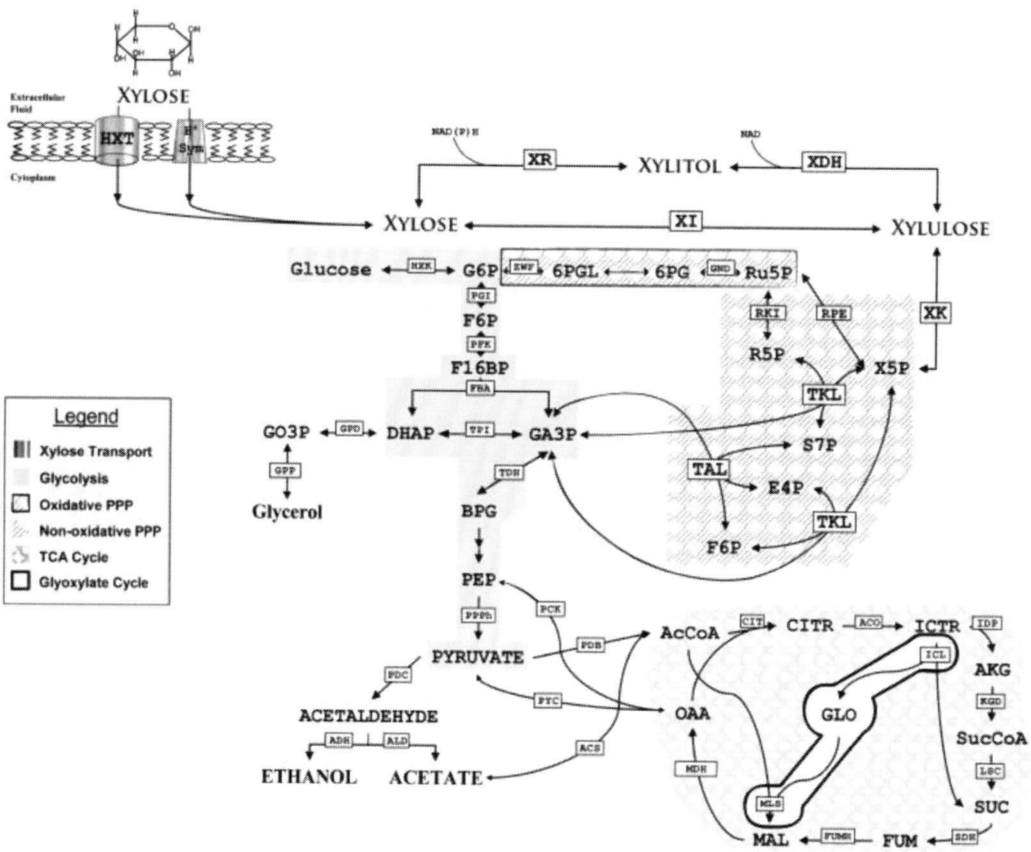

Figure 8. Metabolic pathway for the ethanol production from xylose (*95*).

Bacteria

Ethanol producing bacteria have attracted much attention in recent years because their growth rate is substantially higher than that of the *Saccharomyces* yeast strains. These microorganisms have the potential to play a key role in making the ethanol production more economical (*86*). Table 6 shows some strains of ethanol producing bacteria.

The most promising microorganism is *Zymomonas mobilis,* which converts one mol of hexose into one mol of ethanol. This bacterium is an obligate fermentative organism and uses the Entner-Doudoroff (ED) pathway to degrade sugars to ethanol (Figure 9). The most striking difference between the metabolic pathway used by *Z. mobilis* (ED pathway) and the one used by *S. cerevisiae* (EMP pathway) is the absence of the step catalyzed by the enzyme phosphofructokinase (PFK) in the ED pathway. PFK is a key enzyme of glycolysis. In yeast, this enzyme is tightly regulated. The absence of the PFK enzyme in *Z. mobilis* allows ethanol production in this organism to be decoupled from energy generation (*10*). Thus, in this bacterium cell growth is not required for ethanol production and high cell concentrations are not needed for high ethanol yield. The energetic efficiency of this pathway is one mole of ATP per mol of hexose compared to 2 ATP per mol of hexose in the EMP pathway. This low energy efficiency results in higher yield (up to 97% of the theoretical maximum).

Additionally, this bacterium has high ethanol tolerance (100-120g/l), high sugar uptake and ethanol production rates, a higher optimal production temperature and no requirements for aeration during the fermentation process for optimal ethanol production.

However, there are some disadvantages in the utilization of this bacterium, as follows:

- Its range of fermentable substrates is very narrow (glucose, fructose and sucrose)(*33, 100*). For this reason, this microorganism is not suitable for degradation of lignocellulosic materials due to its narrow range of assimilable substrates.
- When this bacterium is used for fermentation of sugar cane syrup and their sucrose-based media, it occurs the formation of the polysaccharide levan (made up of fructose units) which increases the viscosity of the fermentation broth, and of sorbitol, a product of the fructose reduction that decreases the efficiency of the conversion of sucrose into ethanol (*101-103*). Nevertheless, the simultaneous production of ethanol and levan has been proposed using immobilized cells of *Z. mobilis*, considering that this polysaccharide can be used in the food industry, and in medicine as an immunomodulator, a blood plasma substitute and prolonger of the effect of medical drugs (*104*).
- Although *Z. mobilis* is normally classified as a safe microorganism (*105*), its biomass is not commonly accepted for use as animal feed.

On the other hand, the Gram-negative strain *Zymobacter palmae*, which was isolated by Okamoto et al. (*97*) using a broad range of carbohydrate substrates, is a facultative anaerobe that ferments hexoses, α-linked di- and tri-saccharides, and sugar alcohols (fructose, galactose, glucose, mannose, maltose, melibiose, sucrose, raffinose, mannitol and sorbitol). This strain produces approximately 2 mol of ethanol per mole of glucose without accumulation of byproducts and shows productivities similar to that of *Z. mobilis(106)*.

Thermophilic and saccharolytic clostridia are other important group of ethanol-producing microorganisms. *Clostridium thermohydrosulfuricum*, *Thermoanaerobacter* (formerly Clostridium) *thermossacharolyticum*, and *Clostridium thermocellum* can synthesize up to 2 mol ethanol/mol hexose. Likewise, those bacteria may transform pentoses and aminoacids into ethanol. Having saccharolytic properties, these microorganisms have the ability to grow on a wide variety of non-treated waste. *C. thermocellum* can even directly convert lignocellulosic materials into ethanol (*99*). Nevertheless, the fact that their cultivation can be carried out at high temperatures offers the possibility for a more effective alcohol removal distillation or pervaporation (*33*). In this point, process integration can play a crucial role for the evaluation of the most optimal configurations in order to implement them at industrial level.

Microorganisms like Klebsiella, Citrobacter, Enterobacter, Clostridium, Lactobacillus, Bacillus, Propionibacterium, and Anaerobiospirillum have been reported for glycerol degradation in a fermentative way. However, these microorganisms present diverse problems for their industrial use such as pathogenicity level, strict anaerobic conditions and complex cultivation media. Thus, it is necessary to search anaerobic systems available to metabolize glycerol without pathogenic effects. One alternative is the E. coli strain that is able to use glycerol as carbon source without any external electron receiver. This process is regulated by glycerol dehydrogenase (GldA) and dihydroxyacetone kinase (DHAK) to obtain ethanol, succinate, acetate, and formate (107). This above mentioned process is resumed in Figure 10. Recently, deletions in E. coli were carried out to increase the formate and ethanol yields from glycerol at 10 g/L (80). Thus, using the overexpression of the glycerol dehydrogenase (gldA) and dihydroxyacetone kinase (dhaKLM) genes, an ethanol yield of 95% from glycerol was achieved. Other authors developed a metabolic characterization to evaluate succinate, acetate, formate, lactate, and ethanol yields (108). The highest ethanol yield was achieved with 12.51 g ethanol/g acetate.

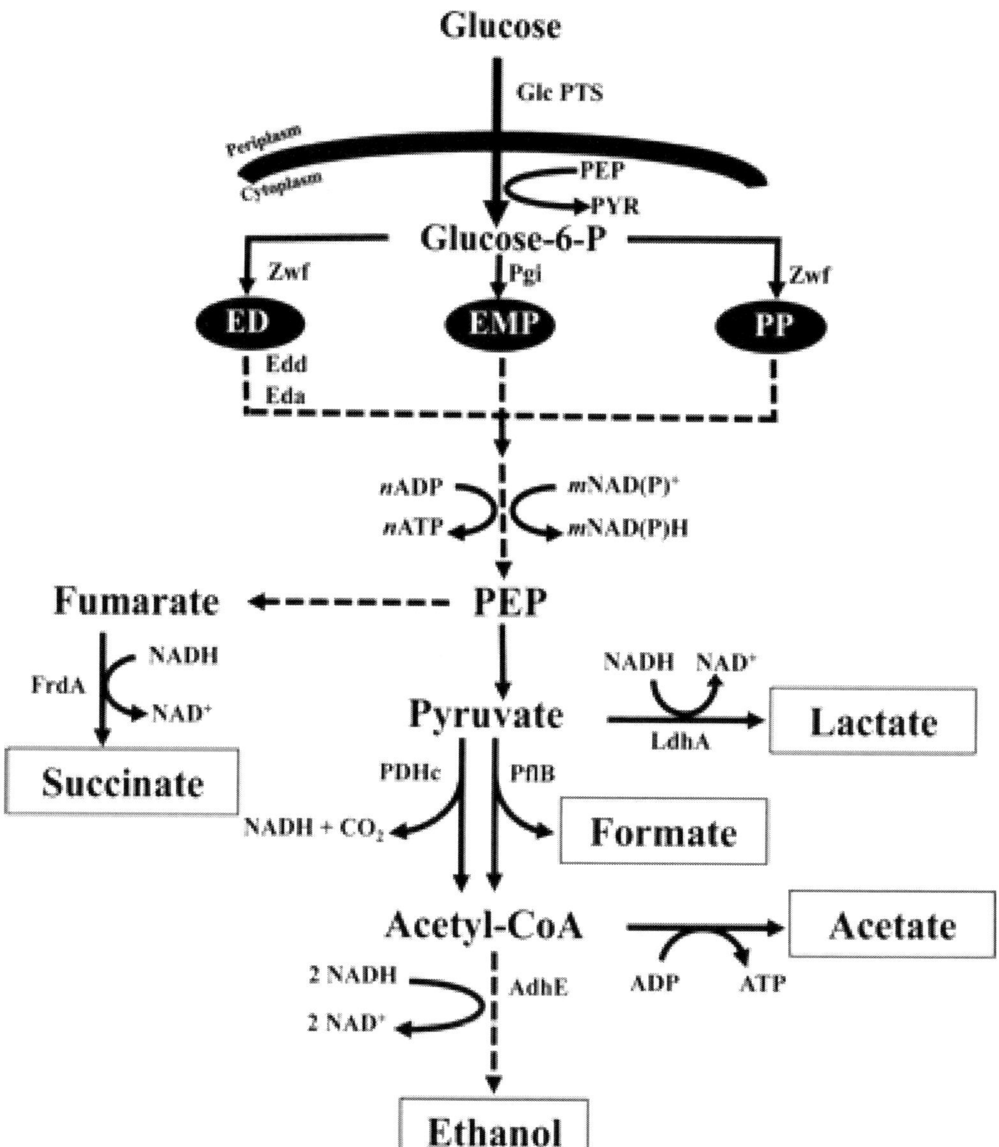

Figure 9.The Entner-Doudoroff (ED) pathway to degrade sugars into ethanol by *Zymomonas mobilis*.

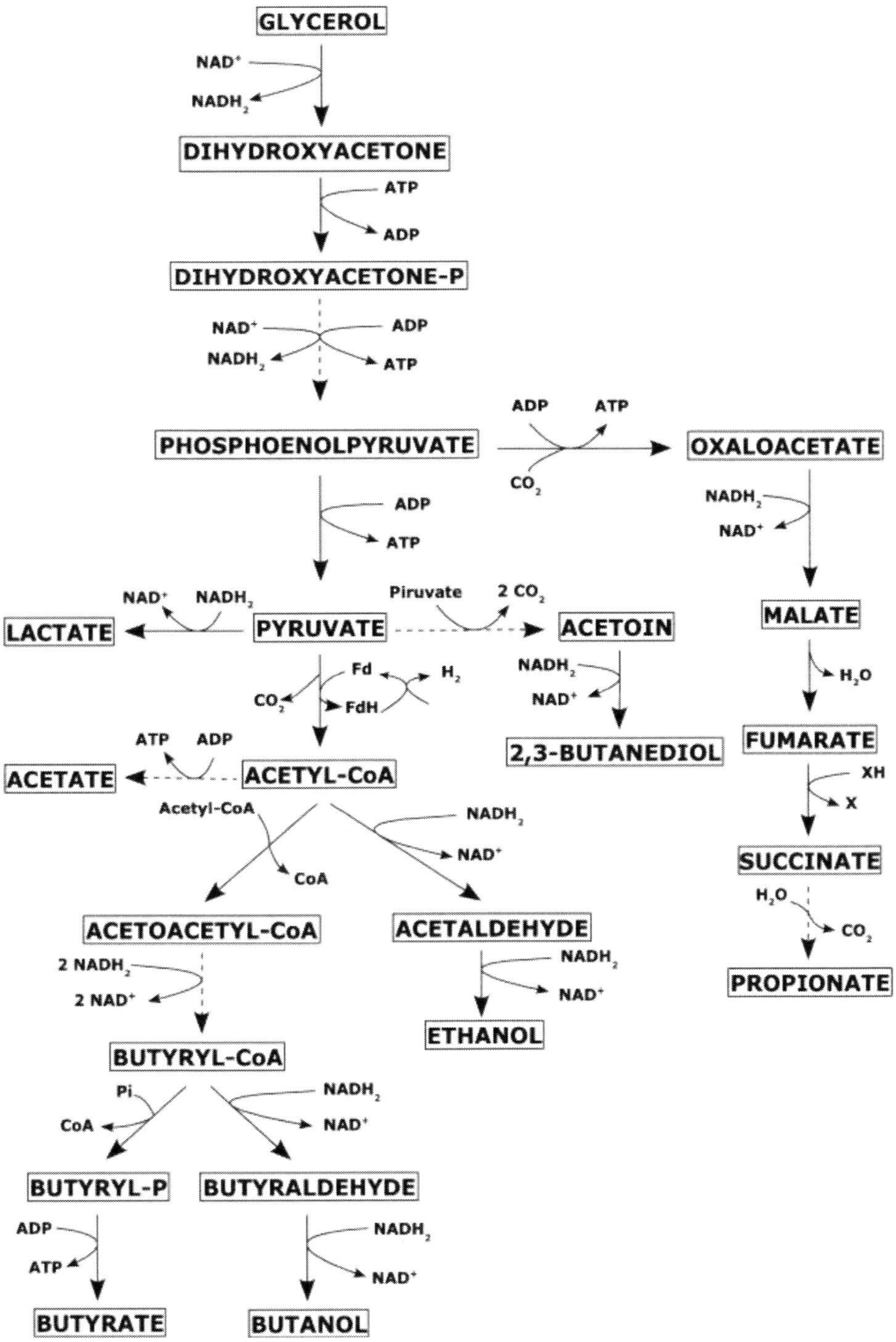

Figure 10. Main metabolic pathways for the fermentative degradation of glycerol by bacteria (*109*).

A mixture of ethanol and formate can be produced by glycerol fermentation using *Klebsiella planticola* isolated from the red deer rumen (*110*). Dharmadi et. al. (*111*)reported the glycerol fermentation by *E. coli,* where pH-dependence and CO_2 availability were analyzed. Ito et al. (*112*) showed that glycerol at 10 g/L was almost completely consumed within 84 h and the main products were ethanol and succinic acid with molar yields of 86% and 7%, respectively. According to these authors, *E. coli* is already a good biocatalyst for glycerol conversion into ethanol and hydrogen. *Enterobacter aerogenes* can be used for ethanol production at high yields from biodiesel wastes containing glycerol. Moreover, a 80 mM glycerol synthetic medium containing biodiesel wastes was analyzed and it was observed that glycerol was consumed in 24 h producing 0.89 mol of H_2 and 1.0 mol of ethanol per mol of glycerol (*112*).

Fungi

Table 5 shows some ethanol producer fungi. The filamentous fungus *Fusarium oxysporum* is known for its ability to produce ethanol by simultaneous saccharification and fermentation (SSF) of cellulose. However, the conversion rate is low and significant amounts of acetic acid are produced as a by-product (*113*). A few microbial species such as *Neurospora*, *Monilia*, *Paecilomyces* and *Fusarium* have been reported to hold the ability to ferment cellulose directly to ethanol (*114*). *F. oxysporum* produces a broad range of cellulases and xylanases, which has been characterised earlier (*115*). Acetic acid was the major fermentation product of *Neocallimastix sp.*, another ethanol producing fungus (*116*).

Although the fermentation stage is the most important step in the ethanol production process since it is in this step where the transformation of substrates into ethanol is made, it is restricted by a series of biological limitations (already mentioned in previous sections of this chapter). These limitations reduce the efficiency and increase the production costs. Hence, many investigations are focused in the improvement of the fermentation stage.

Table 5. Some ethanol producer fungi

Fungus strain	Remarks	Reference
Fusarium oxysporum	Produce ethanol from cellulose but significant amounts of acetic acid are produced too	(*113*)
Neurospora, Monilia, Paecilomyces and Fusarium	Has he ability to ferment cellulose directly to ethanol	(*114*)
Neocallimastix sp.	Ethanol producing fungus but produce acetic acid as the major fermentation product	(*116*)
Trichoderma reesei	Use cellulose directly and is implemented in the SSF process usually	(*34*)

Improvement of the Fermentation Stage in Ethanol Production

The improvement of the fermentation stage can be made from various flanks: microorganisms, operation regime and the technological configuration.

Improvement with Microorganisms:

The microorganisms used for ethanol production from sugars, starch, glycerol or lignocellulosic materials, must comply with the following ideal characteristics:

- Consume economic substrates such as agroindustrial wastes.
- Decrease the pretreatment cost, consuming raw materials in their natural form.
- Reduction of both the inhibition phenomenon and loss of activity during the fermentation.
- Avoid the production of unwanted by-products.
- Decrease the stages and costs for biomass separation.

These characteristics can be obtained by improving alternative strategies with microorganisms. Here we discuss some of them.

Isolation of New Wild Strains

The first level is the search and isolation of wild strains of microorganisms that have the ability to produce ethanol from complex substrates (e.g., disaccharides, cellulose, hemicelluloses), high tolerance to increased working temperature and ethanol concentration and low production of by-products. One of the main advantages when working with wild strains is the genetic stability of the microorganisms. There are more than 6000.000 strains of non-described yeasts and approximately 10% of them are ascomycetes (this group includes the yeasts used in industry). However, the production of ethanol by wild strains has biological limitations and they are controlled by its genome. Therefore, the genome of the microorganism must be genotypically modified to improve the potential ethanol yield.

Genetic Modifications

The different genetic modifications techniques can be classified into two groups: non induced genetic techniques and induced genetic techniques.

Non-Induced Genetic Techniques

This technique uses the natural mutations that occur in a microbial culture. The probability of genetic change in a cellular division is very low. This probability is greater at shorter generation times. This technique can be divided into two stages: the first stage is the adaptation of the strains to new conditions (e.g. media concentrations, temperature, pH). The goal is to drive the evolutionary change through aggressive pressure in the fermentation conditions during long culturing times. In this period, there are many cell divisions and the probability of natural mutations increases. In the second stages, the mutants who have the appropriated characteristics are isolated and purified. One of the advantages of this technique is that it does not require a previous knowledge of the involved metabolic pathway, transport or regulation processes. The main disadvantage is that the addition or removal of genotypic characteristics may affect the general performance of the microorganisms.

Induced Genetic Techniques

Part of the microorganisms' genome could be directly modified by controlled or non-controlled techniques.

The non-controlled techniques may be by extreme environmental conditions or by genetic recombination techniques. Some of them are summarized here:

Non-Controlled Techniques

- *Mutation:* It is a change in the genome sequence. Mutation may be caused in a natural or induced form. According to the mechanisms, the mutation can be: i) in the genome, affecting the number of chromosomes; ii) in the chromosome, affecting the order of a gen sequence; iii) in the gen (specific mutation), affecting the order of base sequences. Induced mutations need mutagenic agents that can be physical (e.g. UV) or chemical.
- *Protoplasts fusion:* Protoplasts are the cells of which cell walls are removed and the cytoplasmic membrane is the outermost layer in such cells. Protoplasts can be obtained by specific lytic enzymes to remove cell wall. Protoplast fusion is a physical phenomenon,during fusion two or more protoplasts come in contact and adhere with one another either spontaneously or in presence of fusion inducing agents. By protoplast fusion it is possible to transfer some useful genes such as disease resistance, nitrogen fixation, rapid growth rate, more product formation rate, protein quality, frost hardiness, drought resistance, herbicide resistance, heat and cold resistance from one species to another. These are powerful techniques for engineering of microbial strains for desirable industrial properties. Morimura et al. (*117*) developed by protoplast fusion and manipulating culture conditions, flocculating strains of *S. cerevisiae* capable of growing at 35ºC and at molasses concentration of 22% (w/v). Under these conditions and using repeated-batch cultures at laboratory scale, ethanol concentration of 91 g/L and productivities of 2.7 g/(L h) were obtained.

Controlled Techniques (Genetic Recombination)

Genetic recombination is a process by which a molecule of nucleic acid (usually DNA, but can also be RNA) is broken and then joined to a different one. Recombination can occur between similar molecules of DNA, as in homologous recombination, or dissimilar molecules, as in non-homologous end joining. Joining different DNAs may be useful to generate new genotypes. The DNA transference may be by:

- *Transformation:* DNA donator is free in the medium
- *Transduction:* DNA transfer is by a virus
- *Conjugation:* It implies the presence of a conjugative plasmid in the donor cell and cell-cell contact.

Operational Regime

Ethanol can be produced by four main types of industrial operations: batch, continuous, fed-batch and semi-continuous. The type of operation used is according to biological characteristics of the microorganisms used, its physicochemical and nutritional requirements, its kinetic, the type of substrate used, and the economical and technological availability.

In batch fermentation, substrate and yeast culture are charged into the bioreactor together with nutrients. Most of the ethanol produced today is done by the batch operation because their investment costs are low, do not require much control and can be accomplished with unskilled labour (*118*). Complete sterilization and management of feedstocks are easier than in the other processes. The other advantage of batch operation is the greater flexibility that can be achieved by using a bioreactor for various product specifications.

The fed-batch operation, which may be regarded as a combination of the batch and continuous operations, is very popular in the ethanol industry. In this operation, the feed solution, which contains the substrate, yeast culture and the required minerals and vitamins, are fed at constant intervals while effluent is removed discontinuously. The main advantage of the fed-batch system is that intermittent feeding of the substrate prevents inhibition and catabolic repression. If the substrate has an inhibitory effect, intermittent addition improves the productivity of the fermentation by maintaining a low substrate concentration.

In the continuous process, feed, which contains the substrate, culture medium and other required nutrients, is pumped continuously into an agitated vessel where the microorganisms are active. The product, which is taken from the top of the bioreactor, contains ethanol, cells, and residual sugar (*119*). It is essential to keep the culture volume constant in a continuous operation, whereas there is volume variation in the fed-batch processes.

In semi-continuous processes, a portion of the culture is withdrawn at intervals and fresh medium is added to the system. There is volume variation in semi-continuous processes. This method has some of the advantages of the continuous and batch operations. There is no need for a separate inoculum vessel, except at the initial start-up. Time is also not wasted in non-productive idle time for cleaning and re-sterilization. Another advantage of this operation is that not much control is required. However, there is a high risk of contamination and mutation due to long cultivation periods and periodic handling. Furthermore, since larger reactor volumes are needed, slightly higher investment costs are required (*118*).

Technological Configurations

The classic configuration employed for fermenting biomass hydrolysates involves a sequential process where the hydrolysis of cellulose and the fermentation are carried out in different units. This configuration is known as separate hydrolysis and fermentation (SHF). In the alternative variant, the simultaneous saccharification and fermentation (SSF), the hydrolysis and fermentation are performed in a single unit. The most employed micro-organism for fermenting lignocellulosic hydrolysates is *S. cerevisiae*, which ferments the hexoses contained in the hydrolysate but not the pentoses.

Separate Hydrolysis and Fermentation (SHF)

When a sequential process is utilized, the solid fraction of the pretreated lignocellulosic material undergoes hydrolysis (saccharification). This fraction contains the cellulose in an accessible form to acids or enzymes. Once hydrolysis is completed, the resulting cellulose hydrolysate is fermented into ethanol. One of the main features of the SHF process is that each step can be performed at its optimal operating conditions. The most important factors to be taken into account for the saccharification step are reaction time, temperature, pH, enzyme dosage and substrate load (*120*).

Simultaneous Saccharification and Fermentation (SSF)

The SSF process has been extensively studied to reduce the inhibition of end products hydrolysis (*121, 122*). In the process, reducing sugars produced in cellulose hydrolysis or saccharification is simultaneously fermented to ethanol, which greatly reduces the product inhibition to the hydrolysis. However, the need of employing more dilute media to reach suitable rheological properties makes that final product concentration be low. In addition, this process operates at non-optimal conditions for hydrolysis and requires higher enzyme dosage, which positively influences on substrate conversion, but negatively on process costs. Considering that enzymes account for an important part of the production costs, it is necessary to find methods that reduce the cellulases doses to be utilized (*120*).

The microorganisms used in the SSF are usually the fungus *Trichoderma reesei* and *S. cerevisiae* (*34*). Hydrolysis is usually the rate limiting process in SSF (*123*). Thermotolerant yeasts and bacteria have been used in the SSF to raise the temperature close to the optimal hydrolysis temperature. Ballesteros et al. (*89*) have identified *Kluyveromyces marxianus* and *K. fragilis* that have the highest ethanol productivity at 42 °C from a number of yeast strains. *K. marxianus* has an ethanol yield of 0.5 g/g cellulose using Solka Floc 200 as substrate at 42 °C. Kadam and Schmidt (*90*) found that the thermotolerant yeast *Candida acidothermophilum*, produced 80% of the theoretical ethanol yield at 40 °C using dilute acid pretreated poplar as substrate. Compared to the two stage hydrolysis–fermentation process, SSF has some advantages like an increase of the hydrolysis rate by conversion of sugars that inhibit the cellulase activity, lower enzyme requirement, higher product yield, lower requirements for sterile conditions since glucose is removed immediately and ethanol is produced, shorter process time and less reactor volume because a single reactor is used. However, ethanol may also exhibit inhibition to the cellulase activity in the SSF process (*34*).

Configurations involving a higher degree of integration as the simultaneous saccharification and fermentation (SSF) have been successfully implemented, especially in the dry-milling process in ethanol production from corn (*124*). The SSF performed at a temperature above 34°C employing a thermotolerant yeast enables the reduction of cooling requirements and the improvement of the conversion process as claimed in the patent of Otto and Escovar-Kousen (*125*). In the wet-milling process, the cascade process is also employed. This process involves the separate realization of saccharification and fermentation where hydrolysis of dextrines by glucoamylase, yeast propagation, pre-fermentation, and fermentation are carried out through a cascade system. *Z. mobilis* also has been researched for ethanol production from dry-milled corn starch. In particular, a system consisting of a continuous packed bed column with immobilized glucoamylase followed by a continuous fluidized bed reactor with immobilized *Z. mobilis* showed an ethanol concentration of 70 g/L and a productivity of 23 g/(L h). In comparison, the continuous SSF carried out in a fluidized bed reactor with co-immobilized glucoamylase and *Z. mobilis* cells showed lower productivity and ethanol concentrations (*126*). New tendencies in corn-to-ethanol industry are aimed at dry-milling processes. The increase in ethanol production capacity in the USA is mainly represented by corn dry-mill ethanol plants (*127*). Other research efforts are oriented to the development of corn hybrids with higher extractable starch or higher fermentable starch content. Genetic engineering can be applied to direct the accumulation of amylases in the endosperm of transgenic corn kernels making possible the utilization of "self-processing" grains (*128*).

According to Sun and Cheng (*34*) this configuration has some disadvantages, which need to be considered for the improvement of SSF, these include incompatible temperature of hydrolysis and fermentation, ethanol tolerance of microbes and inhibition of enzymes by ethanol.

Simultaneous Saccharification and Co-Fermentation

Other promising integration alternative is the inclusion of the pentose fermentation in the SSF process called simultaneous saccharification and co-fermentation (SSCF) (*120*). In an initial stage, the co-fermentation of mixed cultures was studied. For example, the co-culture of *P. stipitis* and *Brettanomyces clausennii* has been utilized for the SSCF of aspen at 38 ºC and pH of 4.8 yielding 369 L EtOH per ton of aspen for 48 h batch process, as reported by Olsson and Hahn-Hägerdal (*63*). In this configuration, it is necessary that both fermenting microorganisms be compatible in terms of operating pH and temperature. Chandrakant and Bisaria (*94*) suggested that a combination of *C. shehatae* and *S. cerevisiae* is suitable for this kind of process. Similarly, a system including the isomerization of xylose and the fermentation with *S. cerevisiae* in a simultaneous way can be utilized. This system has been proven in non-pretreated spent sulfite liquor and in pretreated acid-hydrolyzed wheat straw (*129*).

Among the drawbacks of this configuration are the high by-product formation in the form of CO_2 and xylitol, poor enzyme stability, incompatible pH and optimal temperature (pH of 7.0 and 70 ºC for the isomerization process), and the reversibility of the enzymatic transformation (*94*).

Simultaneous Reaction and Separation

The afore mentioned type of integrations allow the increase of process efficiency through the improvement of reaction processes. However, separation is the step where major costs are generated in process industry. Therefore, a reaction–separation integration could have the highest impact on the overall process in comparison with homogeneous integration of processes (reaction–reaction, separation–separation) (*120, 130*). This has been demonstrated in the case of a reactive distillation processes (*131*), particularly, in the lactic acid recovery by reactive distillation (*132*), and in the production of allyl alcohol by allyl acetate butanolysis (*133, 134*).

The most promising technological configurations for fuel ethanol production are the systems in which the removal of the product that causes the inhibition is accomplished through an extractive biocompatible agent (solvent). Migration of ethanol to the solvent phase is known as extractive fermentation process. Minier and Goma (*135*) showed that primary aliphatic alcohols with a chain length having less than twelve carbons inhibit the growth of yeast cells. They chose the fatty alcohol n-dodecanol as a solvent for *in-situ* extraction of ethanol in an especial continuous pulse packed column with immobilized cells of *S.*

cerevisiae. This configuration allowed the utilization of very concentrated glucose feed due to the reduction of ethanol in the culture broth. In addition, immobilization seems to protect the cells against solvent toxicity (*136*).

Gyamerah and Glover (*137*) implemented a process where the fermentation stage was coupled with an apparatus for liquid–liquid extraction in a continuous regime at pilot-scale level. They chose n-dodecanol by its very low toxicity for ethanol-producing microorganisms. However, this solvent has some drawbacks: it tends to form a stable emulsion with the culture broth, its melting point is relatively high (26 ºC) considering the fermentation conditions, and its distribution coefficient related to water is not very high. Kollerup and Daugulis (*138*) proposed a mathematical model to describe an extractive fermentation process for a continuous production of ethanol from a glucose-containing medium. In this model, a simple relationship between ethanol concentration in the aqueous phase and ethanol content in the solvent phase was considered. Additionally, a kinetic description of microbial growth did not take into account the inhibition effect due to high concentrations of substrate. Gutiérrez et al. (*139*) modeled the batch extractive co-fermentation from pretreated hydrolyzed lignocellulosic biomass coupling the equations representing the kinetics of the biological process with the equations describing the liquid–liquid equilibrium using different activity models. Liquid medium from this bioreactor is continuously removed in order to separate the cells and carry out the decantation of both phases. The aqueous phase is recycled back to the bioreactor and the ethanol-rich solvent phase is flashed for the regeneration of solvent and the production of almost pure ethanol combination of high cell concentrations and high flow rates. Immobilization of cells may also improve their genetic stability (*118*).

Main Perspectives Related to the Bioethanol Production Process

Due to the future perspectives in the ethanol market, its production still represents a challenge for researchers of the entire world in search of economical, environmental, technical and social feasibility ethanol bio- production. Also, efforts must be made to find raw materials that don't compete with food security, more efficient and less energy consuming pretreatment processes and separation processes with low operational and investment costs.

Seeking for non-food feedstock that can be converted in a simple way to fermenting sugars is one of the challenges that must be faced. However, most perspective feedstocks comprises lignocellulosic materials, hence a complex pretreatment for sugars release must be expected. Pretreatment technologies and detoxification of co-lateral inhibitors obtained during pretreatment are other important challenges that must be faced. Several pretreatment and dctoxification technologies have been proposed, but at a higher costs due to expensive pretreatment processes.

Some technologies are gaining importance due to their potential improvement of processes and substantially decreasing the production costs. Some of these most relevant technologies are:

Consolidated Bioprocessing (CBP)

Biomass processing schemes involving enzymatic or microbial hydrolysis commonly involve four biologically mediated transformations: the production of saccharolytic enzymes (cellulases and hemicellulases); the hydrolysis of carbohydrate components present in pretreated biomass to sugars; the fermentation of hexose sugars (glucose, mannose and galactose); and the fermentation of pentose sugars (xylose and arabinose). These four transformations occur in a single step in a process configuration called consolidated bioprocessing (CBP). This process is distinguished from other less highly integrated configurations in that it does not involve a dedicated process step for cellulose production. Thermochemical processing options appear more promising than biological options for the conversion of the lignin fraction of cellulosic biomass, which can have a detrimental effect on enzymatic hydrolysis but also serves as a source of process energy and potential co-products that have important benefits in a life-cycle context (*140*).

Inmovilized Cells

Fermentation systems operated in continuous mode offer a number of advantages compared to batch processes, generally resulting in enhanced volumetric productivity and, consequently, smaller bioreactor volumes and lower investment and operational costs. These continuous processes can benefit from whole cell immobilization techniques in order to retain high cell densities inside the bioreactors. Such immobilization techniques can be divided into four categories: attachment or adsorption to solid surfaces (e.g. wood chips, brewer's spent grains, DEAE cellulose, and porous glass), entrapment within a porous matrix (e.g. calcium alginate, k-carrageenan, polyvinyl alcohol, agar, gelatine, chitosan, and polyacrilamide), mechanical containment behind a barrier (e.g. microporous membrane filters, and micro-capsules) and self-aggregation of the cells by flocculation. Many aspects related to the application of these distinct immobilization methodologies and carriers have been recently discussed, including their impact in microbial growth and physiology, internal and external mass transfer limitations, product quality and consistency, bioreactor design, bioprocess engineering and economics. Industrial application of continuous fermentation systems with immobilized cells has up till now been scarce, mainly due to process engineering and microbial activity problems as well as unrealized cost advantages (*141*).

Ethanol Production from Microalgae

Certain species of microalgae have the ability of producing high levels of carbohydrates instead of lipids as reserve polymers. These species are ideal candidates for the production of bio-ethanol as carbohydrates from microalgae and can be extracted to produce fermentable sugars. It has been estimated that approximately 5000– 15,000 gal of ethanol/acre/year (46,760–140,290 L/ha) can be produced from microalgae. This yield is several orders of magnitude larger than yields obtained for other feedstocks. Even switchgrass, considered as the cellulosic "second generation" of biofuels, achieves ethanol yields that are a fraction of microalgal yield. The microalgae *Chlorella vulgaris*, has been particularly considered as a promising feedstock for bio-ethanol production because it can accumulate up to 37% (dry weight) of starch. However, higher starch contents can also be obtained for optimized culture conditions. Under favorable climate conditions, yields of 80–100 ton dried Chlorella biomass per ha area for a 300-day culture season can be reached. According to some authors, the Chlorella sp. strain is able to accumulate starch up to 70% of algal dry weight under conditions of protein synthesis suppression. *Chlorococum sp.* was also used as a substrate for bioethanol production under different fermentation conditions. Results showed a maximum bio-ethanol concentration of 3.83 g/L obtained from 10 g/L of lipid-extracted microalgae debris. Production of ethanol by using microalgae as raw material can be performed according to the following procedure. In the first step, microalgae cultivation using sunlight energy is carried out in open or covered ponds or closed photobioreactors, based on tubular, flat plate or other designs. In the second step, the biomass needs to be concentrated by an initial factor of at least about thirty-fold, requiring very low-cost harvesting processes. After harvesting, microalgal starch is extracted from the cells with the aid of mechanical equipment or enzymes. Following starch extraction, amylolytic enzymes are used to promote formation of fermentable sugars. S. cerevisiae is then added to begin alcoholic fermentation. At the end of fermentation, fermented broth containing ethanol is drained from the tank and pumped to a holding tank to be fed to a distillation unit (*141*).

Very High Gravity (VHG) Ethanol Fermentation Technologies

High gravity (HG) ethanol fermentation technologies were proposed in the 1980s and successfully applied in the ethanol fermentation industry thereafter, which made the ethanol concentration at the end of fermentation increase dramatically from previously 6-8% (w/v) to currently 10-12% (w/v). Research in yeast physiology has revealed that many strains in the genus of *S. cerevisiae* can tolerate far higher ethanol concentration than assumed; usually without any conditioning or genetic modifications that risk making the modified strains lose some of their original traits. Therefore, very high gravity (VHG) ethanol fermentation technologies using media containing sugar in excess of 250 $g\ l^{-1}$ in order to achieve more than 15% (v/v) ethanol were proposed in the middle of the 1990s (*6*).

REFERENCES

[1] Cardona, C. A., Sánchez, O. J., and Gutiérrez, L. P. (2010) *Process synthesis for fuel ethanol production*.

[2] ASOCAÑA. (2010) Anual Inform 2010-2011, p 54, Colombia.

[3] Quintero, J. A. (2011) Design and Evaluation of Fuel Alcohol Production from Lignocellulosic Raw Materials, In *Departamento de Ingeniería Eléctrica, Electrónica y Computación*, p 177, Universidad Nacional de Colombia Sede Manizales, Manizales, Colombia.

[4] Quintero, J. A., and Cardona, C. A. (2009) Nuevas tecnologías en la produccion de bioetanol a partir de biomasa lignocelulósica, In *Avances investigativos en la producción de biocombustible* (Orrego, C. A. C.-C. E., Ed.), pp 171-198, Manizales, Caldas,Colombia.

[5] Sánchez, O. J., and Cardona, C. A. (2005) Producción biotecnológica de alcohol carburante II: Integración de procesos, *Interciencia 30*, 679-686.

[6] Bai, F. W. (2007) Process Oscillations in Continuous Ethanol Fermentation with Saccharomyces cerevisiae, University of Waterloo, Ontario, Canada.

[7] Cardona, C. A., Quintero, J. A., and Paz, I. C. (2010) Production of bioethanol from sugarcane bagasse: Status and perspectives, *Bioresource Technology 101*, 4754-4766.

[8] Cavalheiro, J. M. B. T., De-Almeida, M. C. M. D., Grandfilis, C., and Da-Fonseca, M. M. R. (2009) Poly(3- hydroxybutyrate) production by Cupriavidus necator using waste glycerol., *Process Biochemistry 44(5)*, 509-515.

[9] Dragone, G., Fernandes, B., Vicente, A. A., and Teixeira, J. A. Third generation biofuels from microalgae, In *Current Research, Technology and Education Topics in Applied Microbiology and Microbial Biotechnology* (Mendez-Vilas, A., Ed.).

[10] Drapcho, C., Nghiem, J., and Walker, T. (2008) *Biofuels engineering process technology.*

[11] Quintero, J., Montoya, M., Sánchez, O., Giraldo, O., and Cardona, C. (2008) Fuel ethanol production from sugarcane and corn: comparative analysis for a colombian case, *Energy 33*, 385-399.

[12] Cardona, C., and Sánchez, O. (2007) *Producción de alcohol carburante: una alternativa para el desarrollo agroindustrial.*

[13] Ogbonna, J., Mashima, H., and Tanaka, H. (2001) Scale up of fuel ethanol production from sugar beet juice using loofa sponge immobilized bioreactor, *Bioresource Technology 76*, 1-8.

[14] Ergun, M., Mutlu, S. F., and Gürel, Ö. (1997) Improved ethanol production by Saccharomyces cerevisiae with EDTA, ferrocyanide and zeolite X addition to sugar beet molasses, *Journal of Chemical Technology and Biotechnology Advances 68*, 147-150.

[15] Castellar, M. R., Aires-Barros, M. R., Cabral, J. M. S., and Iborra, J. L. (1998) Effect of zeolite addition on ethanol production from glucose by Saccharomyces bayanus, *Journal of Chemical Technology and Biotechnology 73*, 377-384.

[16] Acevedo, A., Godoy, R., and Bolaños, G. (2003) Increase in ethanol production during fermentation of molasses using the enzymatic complex Rhyzozyme, In *XXII Congreso Colombiano de Ingeniería Química*, Bucaramanga, Colombia.

[17] Maye, J. P. (2006) Use of hop acids in fuel ethanol production, (US2006263484, Ed.).

[18] Robertson, G. H., Wong, D. W. S., Lee, C. C., Wagschal, K., M.R. Smith, and Orts, W. J. (2006) Native or raw starch digestion: a key step in energy efficient biorefining of grain, *Journal of Agricultural and Food Chemistry 54*, 353-365.

[19] Pandey, A., Nigam, P., Soccol, C. R., Soccol, V. T., Singh, D., and Mohan, R. (2000) Advances in microbial amylases, *Biotechnology and Applied Biochemistry 31*, 135-152.

[20] Shigechi, H., Fujita, Y., Koh, J., Ueda, M., Fukuda, H., and Kondo, A. (2004) Energy-saving direct ethanol production from low-temperaturecooked corn starch using a cell-surface engineered yeast strain codisplaying glucoamylase and a-amylase, *Biochemical Engineering 18*, 149-153.

[21] Wang, S., Ingledew, W., Thomas, K., Sosulski, K., and Sosulski, F. (1999) Optimization of fermentation temperature and mash specific gravity for fuel alcohol production, *Cereal Chemistry 76(1)*, 82-86.

[22] Soni, S. K., Kaur, A., and Gupta, J. K. (2003) A solid state fermentation based bacterial a-amylase and fungal glucoamylase system and its suitability for the hydrolysis of wheat starch, *Process Biochemistry 39*, 185-192.

[23] Barber, A. R., Henningsson, M., and Pamment, N. B. (2002) Aceletarion of high gravity yeast fermentations by acetaldehyde addition, *Renewable Energy 30*, 1733-1747.

[24] FAO. (2004) Global cassava market study. Business opportunities for the use of cassava. Proceedings of the Validation Forum on the Global Cassava Development Strategy, Food and Agriculture Organization of the United Nations (FAO), Rome.

[25] López-Ulibarri, R., and Hall, G. M. (1997) Saccharification of cassava flour starch in a hollow-fiber membrane reactor, *Enzyme and Microbial Technology 21*, 398-404.

[26] Sriroth, K., Lamchaiyaphum, B., and Piyachomkwan, K. (2007) Present situation and future potential of cassava in Thailand.

[27] Wang, S., Sosulski, K., Sosulski, F., and Ingledew, M. (1997) Effect of sequential abrasion on starch composition of five cereals for ethanol fermentation, *Food Research International 30 (8)*, 603–609.

[28] Zhan, X., Wang, D., Tuinstra, M. R., Bean, S., Seib, P. A., and Sun, X. S. (2003) Ethanol and lactic acid production as affected by sorghum genotype and location, *Industrial Crops and Products 18*, 245-255.

[29] Hammond, J. B., Egg, R., Diggins, D., and Coble, C. G. (1996) Alcohol frombananas, *Bioresource Technology 56*, 125-130.

[30] Chung, B. H., and Nam, J. G. (2002) Process for producing high concentration of ethanol using food wastes by fermentation.

[31] Winner-Network. (2002) Village Level Bioenergy System Based on Sweet Sorghum.

[32] Grassi, G. (1999) Modern bioenergy in the European Union. Renewable, *Energy 16*, 985-990.

[33] Claassen, P. A. M., Van-Lier, J. B., López, A. M., Van-Niel, E. W. J., Sijtsma, L., Stams, A. J. M., De-Vries, S. S., and Weusthuis, R. A. (1999) Utilization of biomass for the supply of energy carriers, *Appl Microbiol Biotechnol 52*, 741-755.

[34] Sun, Y., and Cheng, J. (2002) Hydrolysis of lignocellulosic materials for ethanol production: a review., *Bioresource Technology 83*, 1-11.

[35] Laureano-Perez, L. (2005) Understanding factors that limit enzymatic hydrolysis of biomass, *Applied Biochemistry and Biotechnology*, 1081-1099.

[36] Saha, B. C. (2005) Dilute acid pretreatment, enzymatic saccharification and fermentation of wheat straw to ethanol, *Process Biochemitry 40*, 3963-3700.

[37] Bobleter, O. (1994) Hydrothermal degradation of polymers derived from plants, *Progress in Polymer Science 19*, 797-841.

[38] Lynd, L. R. (1996) Overview and evaluation of fuel ethanol from cellulosic biomass: technology, Economics, the Environment, and Policy., *Annual Review of Energy and the Environment 21*, 403–465.

[39] Ghosh, P., and Ghose, T. K. (2003) Bioethanol in India: recent past and emerging future, *Advances in Biochemical Engineering/Biotechnology 85*, 1-27.

[40] Shahbazi, A., Li, Y., and Mims, M. R. (2005) Application of sequential aqueous steam treatments to the fractionation of softwood, *Applied Biochemistry and Biotechnology*, 973-987.

[41] Söderström, J., Pilcher, L., Galbe, M., and Zacchi, G. (2003) Two-step steam pretreatment of softwood by dilute H2SO4 impregnation for ethanol production., *Biomass and Bioenergy 24*, 475-486.

[42] Ballesteros, I., Oliva, J. M., Negro, M. J., Manzanares, P., and Ballesteros, M. (2002a) Enzymic hydrolysis of steam exploded herbaceous agricultural waste (Brassica carinata) at different particule sizes, *Process Biochemistry 38*, 187-192.

[43] Laser, M., Schulman, D., Allen, S. G., Lichwa, J., Antal, J. M. J., and Lynd, L. R. (2002) A comparison of liquid hot water and steam pretreatments of sugar cane bagasse for bioconversion to ethanol, *Bioresource Technology 81*, 33-44.

[44] Ogier, J. C., Ballerini, D., Leygue, J. P., Rigal, L., and Pourquié, J. (1999) Production d'éthanol à partir de biomasse lignocellulosique, *Oil & Gas Science and Technology – Revue de l'IFP (in french) 54*, 67-94.

[45] Aden, A., Ruth, M., Ibsen, K., Jechura, J., Neeves, K., Sheehan, J., Wallace, B., Montague, L., Slayton, A., and Lukas, J. (2002) Lignocellulosic biomass to ethanol process design and economics utilizing cocurrent dilute acid prehydrolysis and enzymatic hydrolysis for corn stover. In *Technical Report NREL/TP-510-32438. Golden, CO (USA),* (Laboratory, N. R. E., Ed.), p 143p.

[46] Wooley, R., Ruth, M., Sheehan, J., Ibsen, K., Majdeski, H., and Gálvez, A. (1999) Lignocellulosic biomass to ethanol process design and economics utilizing co-current dilute acid prehydrolysis and enzymatic hydrolysis. Current and futuristic scenarios, In *Technical Report NREL/ TP-580-26157. Golden, CO (USA)* (Laboratory, N. R. E., Ed.), p 123p.

[47] Jacobsen, S. E., and Wyman, C. E. (2002) Xylose monomer and oligomer yields for uncatalyzed hydrolysis of sugarcane bagasse hemicellulose at varying solids concentration, *Industrial & Engineering Chemistry Research 41*, 1454–1461.

[48] Saha, B. C., Iten, L. B., Cotta, M. A., and Wu, Y. V. (2005) Dilute acid pretreatment, enzymatic saccharification and fermentation of wheat straw to ethanol, *Process Biochemistry 40*, 3693–3700.

[49] Farone, W. A., and Cuzens, J. E. (1996) Method of strong acid hydrolysis.

[50] Cuzens, J. C., and Miller, J. R. (1997) Acid hydrolysis of bagasse for ethanol production, *Renewable Energy 10 (2/3)*, 285–290.

[51] Lynd, L. R., Wyman, C. E., and Gerngross, T. U. (1999) Biocommodity engineering, *Biotechnology Progress 15*, 777-793.

[52] Zhang, B. S. (2006) Process for preparing fuel ethanol by using straw fiber materials.

[53] Tengerdy, R. P., and Szakacs, G. (2003) Bioconversion of lignocellulose in solid substrate fermentation, *Biochemical Engineering Journal 13*, 169-179.

[54] Kang, S. W., Park, Y. S., Lee, J. S., Hong, S. I., and Kim, S. W. (2004) Production of cellulases and hemicellulases by Aspergillus niger KK2 from lignocellulosic biomass, *Bioresource Technology 91*, 153-156.

[55] Hamelinck, C. N., Van-Hooijdonk, G., and Faaij, A. P. C. (2005) Ethanol from lignocellulosic biomass: techno-economic performance in short-, middle-and long-term., *Biomass and Bioenergy 28*, 384-410.

[56] Ojumu, T. V., and Ogunkunle, O. A. (2005) Production of glucose from lignocellulosic under extremely low acid and high temperature in batch process, auto-hydrolysis approach, *Journal of Applied Sciences 5(1)*, 15-17.

[57] Zhang, Y. H. P., and Lynd, L. R. (2004) Toward an aggregated understanding of enzymatic hydrolysis of cellulose: noncomplexed cellulose systems., *Biotechnology and Bioengineering 8(7)*, 797-824.

[58] Lynd, L. R., Weimer, P. J., Van-Zyl, W. H., and Pretorious, I. S. (2002) Microbial cellulose utilization: fundamentals and biotechnology., *Microbiology and Molecular Biology Reviews 66 (3)*, 506–577.

[59] Kim, S., and Dale, B. E. (2004) Global potential bioethanol production from wasted crops and crop residues, *Biomass and Bioenergy 26*, 361-375.

[60] Palmqvist, E., and Hahn-Hägerdal, B. (2000) Fermentation of lignocellulosic hydrolysates. II: inhibitors and mechanisms of inhibition, *Bioresource Technology 74*, 25-33.

[61] Palmqvist, E., and Hahn-Hägerdal, B. (2000) Fermentation of lignocellulosic hydrolysates. I: inhibition and detoxification, *Bioresource Technology 74*, 17-24.

[62] Persson, P., Andersson, J., Gorton, L., Larsson, S., Nilvebrant, N. O., and Jönsson, L. F. (2002) Effect of different forms of alkali treatment on specific fermentation inhibitors and on the fermentability of lignocellulose hydrolysates for production of fuel ethanol, *Journal of Agricultural and Food Chemistry 50*, 5318-5325.

[63] Olsson, L., and Hahn-Hägerdal, B. (1996.) Fermentation of lignocellulosic hydrolysates for ethanol production, *Enzyme and Microbial Technology 18*, 312-331.

[64] Yu, Z., and Zhang, H. (2003) Pretreatments of cellulose pyrolysate for ethanol production by Saccharomyces cerevisiae, Pichia sp. YZ-1 and Zymomonas mobilis, *Biomass and Bioenergy 24*, 257-262.

[65] Khiyami, M. A., Pometto-III, A. L., and Brown, R. C. (2005) Detoxification of corn stover and corn starch pyrolysis liquors by Pseudomonas putida and Streptomyces setonii suspended cells and plastic compost support biofilms, *Journal of Agricultural and Food Chemistry 53*, 2978-2987.

[66] Persson, P., Larsson, S., Jönsson, L. J., Nilvebrant, N. O., Sivik, B., Munteanu, F., Thörneby, L., and Gorton, L. (2002) Supercritical fluid extraction of a lignocellulosic hydrolysate of spruce for detoxification and to facilitate analysis of inhibitors. *Biotechnology and Bioengineering 79*, 694-700.

[67] Taherzadeh, M. J. (1999) Ethanol from Lignocellulose: Physiological Effects of Inhibitors and Fermentation Strategies, p 56, Chalmers, University of Technology Göteborg Sweden.

[68] Nilsson, A., Taherzadeh, M. J., and Lidén, G. (2001) Use of dynamic step response for control of fed-batch conversion of lignocellulosic hydrolyzates to ethanol, *Journal of Biotechnology 89*, 41-43.

[69] Purwadi, R., Brandberg, T., and Taherzadeh, M. J. (2007) A possible industrial solution to ferment lignocellulosic hydrolyzate to ethanol: continuous cultivation with flocculating yeast. *International Journal of Molecular Sciences 8*, 920-932.

[70] Palmqvist, E., Grage, H., Meinander, N. Q., and Hahn-Hägerdal, B. (1999) Main and interaction effects of acetic acid, furfural, and p-hydroxybenzoic acid on growth and ethanol productivity of yeasts. *Biotechnology and Bioengineering 63 (1)*, 46-55.

[71] Oliva, J. M., Ballesteros, I., Negro, M. J., Manzanares, P., Cabañas, A., and Ballesteros, M. (2004) Effect of binary combinations of selected toxic compounds on growth and fermentation of Kluyveromyces marxianus, *Biotechnology Progress 20*, 715-720.

[72] Oliva, J. M., Sáez, F., Ballesteros, I., González, A., Negro, M. J., Manzanares, P., and Ballesteros, M. (2003) Effect of lignocellulosic degradation compounds from steam explosion pretreatment on ethanol fermentation by thermotolerant yeast Kluyveromyces marxianus., *Applied Biochemistry and Biotechnology 2003*, 141-153.

[73] Zaldivar, J., Martinez, A., and Ingram, L. O. (1999) Effect of selected aldehydes on the growth and fermentation of ethanologenic Escherichia coli., *Biotechnology and Bioengineering 65*, 24-33.

[74] Zaldivar, J., Martinez, A., and Ingram, L. O. (2000) Effect of alcohol compounds found in hemicellulose hydrolysate on the growth and fermentation of ethanologenic Escherichia coli, *Biotechnology and Bioengineering 68(5)*, 524-530.

[75] Belkacemi, K., Turcotte, G., and Savoie, P. (2002) Aqueous/steam-fractionated agricultural residues as substrates for ethanol production, *Industrial & Engineering Chemistry Research 41*, 173-179.

[76] Wang, Z. X., Zhuge, J., Fang, H., and Prior, B. A. (2001) Glycerol production by microbial fermentation: a review., *Biotechnol Adv 19*, 201-223.

[77] Solomon, B. O., Zeng, A. P., Biebl, H., Schlieker, H., Posten, C., and Deckwer, W. D. (1995) Comparison of the energetic efficiencies of hydrogen and oxychemicals formation in Klebsiella pneumoniae and Clostridium butyricum during anaerobic growth on glycerol., *J Biotechnol 39*, 107-117.

[78] Barbirato, F., and Bories, A. (1997) Relationship between the physiology of Enterobacter agglomerans CNCM 1210 grown anaerobically on glycerol and the culture conditions., *Res Microbiol 148*, 475-484.

[79] Menzel, K., Zeng, A. P., and Deckwer, W. D. (1997) High concentration and productivity of 1,3-propanediol from continuous fermentation of glycerol by Klebsiella pneumoniae, *Enzym Microb Technol 20*, 82-86.

[80] Yazdani, S. S., and A., R. G. (2007) Anaerobic fermentation of glycerol: a path to economic viability for the biofuels industry, *Curr Opin Biotechnol 18*, 213-219.

[81] Ligthelm, M. E., Prior, B. A., and du-Preez, J. C. (1988) The oxygen requirements of yeasts for the fermentation of d-xylose and d-glucose to ethanol, *Applied Microbiology and Biotechnology 28*, 63-68.

[82] Skoog, K., Jeppsson, H., and Hahn-Hagerdal, B. (1992) The effect of oxygenation on glucose fermentation with Pichia stipitis, *Applied Microbiology and Biotechnology 34-35*, 369-375.

[83] Ballesteros, I., Oliva, J. M., Sáez, F., and Ballesteros, M. (2001) Ethanol production from lignocellulosic byproducts of olive oil extraction, *Applied Biochemistry and Biotechnology 91*, 237-252.

[84] Hahn-Hägerdal, B., Jeppsson, H., Olsson, L., and Mohagheghi, A. (1994) An interlaboratory comparison of the performance of ethanol-producing micro-organisms in a xylose-rich acid hydroysate, *Applied Microbiology and Biotechnology 41*, 62-72.

[85] van Zyl, C., Prior, B. A., and du Preez, J. C. (1991) Acetic acid inhibition of d-xylose fermentation by Pichia stipitis, *Enzyme and Microbial Technology 13*, 82-86.

[86] Dien, B. S., Cotta, M. A., and Jeffries, T. W. (2003) Bacteria engineered for fuel ethanol production: current status, *Applied Microbiology and Biotechnology 63*, 258-266.

[87] Olsson, L., Linden, T., and Hahn-Hagerdal, B. (1992) Performance of microorganisms in spent sulfite liquor and enzymatic hydrolysate of steam-pretreated Salix, *Applied Biochemistry and Biotechnology 34-35*, 359-367.

[88] Olsson, L., and Hahn-Hägerdal, B. (1993) Fermentative performance of bacteria and yeasts in lignocellulose hydrolysates, *Process Biochemistry 28*, 249-257.

[89] Ballesteros, I., Ballesteros, M., Cabañas, A., Carrasco, J., MartÍn, C., Negro, M. J., Saez, F., and Saez, R. (1991) Selection of thermotolerant yeasts for simultaneous saccharification and fermentation (SSF) of cellulose to ethanol, *Applied Biochemistry and Biotechnology 28-29*, 307-315.

[90] Kadam, K. L., and Schmidt, S. L. (1997) Evaluation of Candida acidothermophilum in ethanol production from lignocellulosic biomass, *Applied Microbiology and Biotechnology 48*, 709-713.

[91] Olsson, L., Lindén, T., and Hahn-HäGerdal, B. (1992) Performance of microorganisms in spent sulfite liquor and enzymatic hydrolysate of steam-pretreated < i> Salix</i&gt, *Applied Biochemistry and Biotechnology 34-35*, 359-368.

[92] Bai, F. W., Anderson, W. A., and Moo-Young, M. (2008) Ethanol fermentation technologies from sugar and starch feedstocks, *Biotechnology Advances 26*, 89-105.

[93] Saithong, P., Nakamura, T., and Shima, J. (2009) Prevention of bacterial contamination using acetate-tolerant Schizosaccharomyces pombe during bioethanol production from molasses, *Journal of Bioscience and Bioengineering 108*, 216-219.

[94] Chandrakant, P., and Bisaria, V. S. (1998) Simultaneous bioconversion of cellulose and hemicellulose to ethanol, *Critical Reviews in Biotechnology 18 (4).*

[95] Chu, B. C. H., and Lee, H. (2007) Genetic improvement of Saccharomyces cerevisiae for xylose fermentation, *Biotechnology Advances 25*, 425-441.

[96] Skotnicki, M. L., Warr, R. G., Goodman, A. E., Lee, K. J., and Rogers, P. L. (1982) High-productivity alcohol fermentations using Zymomonas mobilis, *Biochemical Society Symposia 19*, 53-86.

[97] Okamoto, T., Taguchi, H., Nakamura, K., Ikenaga, H., Kuraishi, H., and Yamasato, K. T. (1993) Zymobacter palmae gen. nov., sp. nov., a new ethanol-fermenting Peritrichous bacterium isolated from palm sap, *Archives of Microbiology 21*, 333-337.

[98] Park, E. Y., Michinaka, A., and Okuda, N. (2001) Enzymatic hydrolysis of waste office paper using viscosity as operating parameter, *Biotechnology Progress 17*, 379-382.

[99] McMillan, J. D. (1997) Bioethanol production: status and prospects, *Renewable Energy 10*, 295-302.

[100] Hawgood, N., Evans, S., and Greenfield, P. F. (1985) Enhanced ethanol production in multiple batch fermentations with an auto-flocculating yeast strain, *Biomass 7*, 261-278.

[101] Doelle, M. B., and Doelle, W. (1989) Ethanol production from sugar cane syrup using Zymomonas mobilis, *Journal of Biotechnoloy 11*, 25-30.

[102] Grote, W., and Rogers, P. L. (1985) Ethanol production from sucrose-based raw materials using inmobilized cells of Zymomonas mobilis, *Biomass 8*, 169-184.

[103] Lee, W. C., and Huang, C. T. (2000) Modeling of ethanol fermentation using Zymomonas mobilis ATCC 10988 grown on the media containing glucose and fructose, *Biochemical Engineering Journal 4*, 217-227.

[104] Karsakevich, A., entina, E., Vlna, E., Bekers, M., Gonta, S., Linde, R., and Kaminska, K. (1998) The effect of chemical treatment of stainless steel wire surfaces on Zymomonas mobilis cell attachment and product synthesis, *Acta Biotechnol 18 (39*, 255-265.

[105] Lin, Y., and S.Tanaka. (2006) Ethanol fermentation from biomass resources: current state and prospects, *Applied Microbiology and Biotechnology 69*, 657-642.

[106] Okamoto, T., (Kirin-Brewery-Co.-Ltd.-Yokohama-(Japan)), Taguchi, H., Nakamura, K., and Ikenaga, H. (1994) Production of ethanol from maltose by Zymobacter palmae fermentation, *Agricultural and Biological Chemistry 58*, 1328-1329.

[107] Murarka, A., Dharmadi, Y., Yazdani, S. S., and González, R. (2008) Fermentative utilization of glycerol by Escherichia coli and its implications for the production of fuels and chemicals., *Appl Environ Microbiol 74*, 1124-1135.

[108] Nikel, P. I., Pettinari, M. J., M.A.Galvagno, and Méndez, B. S. (2008) Poly(3-hydroxybutyrate) synthesis from glycerol by a recombinant Escherichia coli arcA mutant in fed-batch microaerobic cultures, *Appl Microbiol Biotechnol 77*, 1337-1343.

[109] Da-Silva, G. P., Mackb, M., and Contiero, J. (2009) Glycerol: A promising and abundant carbon source for industrial microbiology, *Biotechnology Advances 27*, 30-39.

[110] Jarvis, G. N., Moore, E. R. B., and Thiele, J. H. (1997) Formate and ethanol are the major products of glycerol fermentation produced by a Klebsiella planticola strain isolated from red deer, *J Appl Microbiol 83*, 166-174.

[111] Dharmadi, Y., Murarka, A., and Gonzalez, R. (2006) Anaerobic fermentation of glycerol by Escherichia coli: a new platform for metabolic engineering., *Biotechnol Bioeng 94*, 821-829.

[112] Ito, T., Y. Nakashimada Y, Senba, K., Matsui, T., and Nishio, N. (2005) Hydrogen and ethanol production from glycerol-containing wastes discharged after biodiesel manufacturing process., *J Biosci Bioeng 100*, 260-265.

[113] Panagiotou, G., Villas-Bôas, S. G., Christakopoulos, P., Nielsen, J., and Olsson, L. (2005) Intracellular metabolite profiling of Fusarium oxysporum converting glucose to ethanol, *Journal of Biotechnology 115*, 425-434.

[114] Singh, A., Kumar, P., and Schügerl, K. (1992) Bioconversion of cellulosic materials to ethanol by filamentous fungi enzymes and Products from Bacteria Fungi and Plant Cells, pp 29-55, Springer Berlin / Heidelberg.

[115] Christakopoulos, P., D, D. K., Macris, B. J., Claeyssens, M., and Bhat, M. K. (1996) Purification and characterization of a major xylanase with cellulase and transferase activities from Fusarium oxysporum, *Carbohydrate Research 289*, 91-104.

[116] Dijkerman, R., Ledeboer, J., Op den Camp, H. J. M., Prins, R. A., and van der Drift, C. (1997) The Anaerobic Fungus <i>Neocallimastix</i> sp. Strain L2: Growth and Production of (Hemi)Cellulolytic Enzymes on a Range of Carbohydrate Substrates, *Current Microbiology 34*, 91-96.

[117] Morimura, S., Zhong, Y. L., and Kida, K. (1997) Ethanol production by repeated-batch fermentation at high temperature in a molasses medium containing a high concentration of total sugar by a thermotolerant flocculating yeast with improved salt-tolerance, *Journal of Fermentation and Bioengineering 83*, 271-274.

[118] Caylak, B., and Vardar-Sukan, F. (1998) Comparison of different production processes for bioethanol, *Turkish Journal of Chemistry 22*, 351-359.

[119] Maiorella, B., Wilke, C. R., and Blanch, H. W. (1981) Alcohol production and recovery, *Advances in Biochemical Engineering/Biotechnology 20*, 9-31.

[120] Cardona, C. A., and Sánchez, O. J. (2007) Fuel ethanol production: Process design trends and integration opportunities, *Bioresource Technology 98*, 2415-2457.

[121] Zheng, Y., Lin, H. M., and Tsao, G. T. (1998) Pretreatment for cellulose hydrolysis by carbon dioxide explosion, *Biotechnology Progress 14*, 890-896.

[122] Saxena, A., Garg, S. K., and Verma, J. (1992) Simultaneous saccharification and fermentation of waste newspaper to ethanol, *Bioresource Technology 39*, 13-15.

[123] Philippidis, G. P., and Smith, T. K. (1995) Limiting factors in the simultaneous saccharification and fermentation process for conversion of cellulosic biomass to fuel ethanol. *Applied Biochemistry and Biotechnology 51-52*, 117-124.
[124] Cardona, C. A., Sánchez, O. J., Montoya, M. I., and Quintero, J. A. (2005) Analysis of fuel ethanol production processes using lignocellulosic biomass and starch as feedstocks, In *Seventh World Congress of Chemical Engineering*, Glasgow, Scotland, UK.
[125] Otto, E., and Escovar-Kousen, J. (2004) Ethanol production by simultaneous saccharification and fermentation (SSF).
[126] Krishnan, M. S., Nghiem, N. P., and Davison, B. H. (1999) Ethanol production from corn starch in a fluidized-bed bioreactor, *Applied Biochemistry and Biotechnology 77-79*, 359-372.
[127] Tiffany, D. G., and Eidman, V. R. (2003) *Factors associated with success of fuel ethanol producers*, USA.
[128] Bothast, R. J., and Schlicher, M. A. (2005) Biotechnological processes for conversion of corn into ethanol, *Applied Microbiology and Biotechnology 67*, 19-25.
[129] Linde̒n, T., and Hahn-Hägerdal, B. (1989) Fermentation of lignocellulose hydrolysates with yeasts and xylose isomerase, *Enzyme and Microbial Technology 11*, 583-589.
[130] Cardona, C. A., Matallana, L. G., and Gutiérrez, L. F. (2009) Environmental performance of reactive separation processes, In *8th World Congress of Chemical Engineering*, Montreal, Canada.
[131] Pisarenko, Y. A., Serafimov, L. A., Cardona, C. A., Efremov, D. L., and Shuwalov, A. S. (2001) Reactive distillation design: analysis of the process statics, *Reviews in Chemical Engineering 17(4)*, 253-327.
[132] Cardona, C. A., López, L. J., and López, F. E. (2004) Lactic acid separation by reactive distillation, *Revista EAFIT (In spanish) 40(135)*, 40-53.
[133] Anokhina, E. A., and al., e. (1996) Basic design stages of combined processes by the example of the production of allyl alcohol by allyl acetate butanolysis. Part 1: the selection of a combined process for the production of allyl alcohol, *Chemical Industry in Russia 9*, 537.
[134] Anokhina, E. A., and et.-al. (1996) Basic design stages of combined processes by the example of the production of allyl alcohol by allyl acetate butanolysis. Part 2: separation of a reaction mixture removed from the reaction-rectification column for allyl alcohol synthesis, *Chemical Industry in Russia 11*, 689.
[135] Minier, M., and Goma, G. (1982) Ethanol production by extractive fermentation, *Biotechnology and Bioengineering 24*, 1565-1579.
[136] Aires-Barros, M. R., Cabral, J. M. S., and Novais, J. M. (1987) Production of ethanol by immobilized Saccharomyces bayanus in an extractive fermentation system, *Biotechnology and Bioengineering 29*, 1097-1104.
[137] Gyamerah, M., and Glover, J. (1996) Production of ethanol by continuous fermentation and liquid–liquid extraction, *Journal of Chemical Technology and Biotechnology 66*, 145-152.
[138] Kollerup, F., and Daugulis, A. J. (1985) A mathematical model for ethanol production by extractive fermentation in a continuous stirred tank fermentor, *Biotechnology and Bioengineering 27*, 1335-1346.

[139] Gutiérrez, L. F., Sánchez, O. J., and Cardona, C. A. (2005) Modeling of batch extractive fermentation for the fuel ethanol production, In *Eighth Conference on Process Integration, Modelling and Optimisation for Energy Saving and Pollution Reduction, PRES 2005*, Giardini Naxos, Italy.

[140] Lynd, L. R., Van-Zyl, W. H., McBride, J. E., and Laser, M. (2005) Consolidated bioprocessing of cellulosic biomass: an update, *Current Opinion in Biotechnology 16*, 577-583.

[141] Mussatto, I. S., Dragone, G., Guimarães, P. M. R., Silva, J. P. A., Carneiro, L. M., Roberto, I. C., A.Vicente, Domingues, L., and Teixeira, J. A. (2010) Technological trends, global market, and challenges of bio-ethanol production, *Biotechnology Advances 28*, 817-830.

In: Ethanol: Production, Cellular Mechanisms and Health Impact ISBN: 978-1-62257-297-7
Editors: Roy I. Henry and Billy P. Woods

Chapter 2

PRODUCTION OF ETHANOL FROM SYNTHESIS GAS IN FLOWED VAPOR-SOLID CATALYTIC SYSTEM: A KEY STEP FOR CONVERTING ALL KINDS OF ORGANIC RESOURCES TO ETHANOL VIA CHEMICAL PROCESS

Yanyong Liu** and Kazuhisa Murata
Energy Technology Research Institute, National Institute of Advanced Industrial Science and Technology, Higashi, Tsukuba, Ibaraki, Japan

ABSTRACT

A chemical process using synthesis gas (so-called as syngas, a mixed gas of CO and H_2) as a platform can produce ethanol from all kinds of organic resources (such as biomass, coal, and so on). The process includes the first gasification step for producing syngas from organic resources and the second liquefaction step for converting syngas to ethanol. The gasification is a thermochemical reaction and the liquefaction is a catalytic reaction. Thus the efficiency of the total process is determined by the performance of the catalysts for converting syngas to ethanol. We investigated three types of catalysts (Cu-based, Rh-based, and MoS-based) for the synthesis of alcohols (including ethanol) from syngas in this study. Firstly, we developed a highly active Cu-based catalyst Cs-Ni-Cu/$Ce_{0.8}Zr_{0.2}O_2$ for producing mixed alcohols from syngas at a reaction temperature of 573 K under a reaction pressure of 3.0 MPa. Both the CO conversion and the selectivity for alcohols over Cs-Ni-Cu/$Ce_{0.8}Zr_{0.2}O_2$ were higher than those over the reported Cu-based catalysts. The selectivity for alcohols was very high (>80%) but the fraction of ethanol in total alcohols was not high over Cs-Ni-Cu/$Ce_{0.8}Zr_{0.2}O_2$ because it mainly formed methanol and iso-butanol. Secondly, we developed a highly active Rh-based catalyst Li-Fe-Rh/$Ce_{0.8}Zr_{0.2}O_2$ for producing ethanol from syngas at 548 K under 2.4 MPa. Both the CO conversion and the selectivity for ethanol over Li-Fe-Rh/$Ce_{0.8}Zr_{0.2}O_2$ were higher than those over the reported Rh-based catalysts. Li-Fe-Rh/$Ce_{0.8}Zr_{0.2}O_2$ selectively formed ethanol among various alcohols although the selectivity for by-product

* Tel/fax: +81-29-861-4776. E-mail: yy.ryuu@aist.go.jp (Y. Liu).

methane was relatively high (>20%). Li-Fe-Rh/$Ce_{0.8}Zr_{0.2}O_2$ had a good catalytic stability in a clean syngas but lost activity gradually in a syngas containing 10 ppm H_2S due to forming inactive RhS phase. Thirdly, we developed a highly active Mo-based catalyst K-Fe-Ni-MoS/Al_{13}-Mont for producing mixed alcohols from syngas at 543 K under 6 MPa. Both the CO conversion and the selectivity for ethanol over K-Fe-Ni-MoS/Al_{13}-Mont were higher than those over the reported MoS-based catalysts. K-Fe-Ni-MoS/Al_{13}-Mont formed mixed alcohols as the main oxygenate compounds. The fraction of ethanol in the product over K-Fe-Ni-MoS/Al_{13}-Mont was higher than that over Cs-Ni-Cu/$Ce_{0.8}Zr_{0.2}O_2$ but was lower than that over Li-Fe-Rh/$Ce_{0.8}Zr_{0.2}O_2$. K-Fe-Ni-MoS/Al_{13}-Mont lost activity gradually in a clean syngas due to losing sulfur from catalyst surface during the reaction but it had a good catalytic stability in a syngas containing 10 ppm H_2S.

Keywords: Ethanol, mixed alcohols, synthesis gas (syngas), Cu-based catalyst, Rh-based catalyst, MoS-based catalyst, alkali promoter, F-T element

INTRODUCTION

In recent years, the development of alternative fuels from renewable resources (such as biomass) to substitute fossil fuels (such as coal, crude oil) has become increasingly important because of global climate change, depletion of fossil fuel resources, and rising crude oil prices [1, 2]. Ethanol has been considered as one of the most important synthetic alternative fuels for the automobiles [3]. And in the U. S., most ethanol is sold as a blend of 10% ethanol and 90% gasoline. In addition to the potential application as a transportation fuel, ethanol also has been used as a potential source of hydrogen for fuel cells, and as a feedstock for the synthesis of variety of chemicals, fuels, and polymers [4, 5].

Worldwide ethanol production in 2005 exceeded 12 billion gallons. The U. S. and Brazil were the largest ethanol producers in the world, each contributing over 4.2 billon gallons in 2005. A major portion of the ethanol produced in the U. S. was used for blending with gasoline, but this mixture replaced only about 2% of all gasoline sold. The Energy Policy Act (EPACT) of 2005 required U. S. fuel ethanol production to increase to 7.5 billion gallons by 2012 [6]. This has promoted a significant increase in the research and development (RandD) effort for ethanol production. Currently, there are three major processes for the large-scale ethanol production in the world: a petrochemical process of ethylene hydration, a biological fermentation process through a sugar platform, and a chemical process through a synthesis gas (so-called as syngas, a mixed gas of CO and H_2) platform [7, 8].

In the petrochemical industry, ethanol is produced from the catalytic hydration of ethylene by an acidic catalyst (as shown in route 1).

$$C_2H_4 + H_2O \rightarrow C_2H_5OH \qquad \text{(Route 1)}$$

Because ethylene is mainly produced from the steam refinery of naphtha, the use of obtained ethanol as a motor fuel can not reduce the CO_2 emission from the view of environmental. Further, the ethylene hydration route is depended on imported oil and thus it is limited by the rising crude oils prices. Hence, it can be predicted that the fraction of ethanol produced by the route of ethylene hydration is difficult to increase in the future.

At the present time, ethanol is mainly produced by a traditional fermentation technology from the edible biomass materials (such as carbohydrates, cornstarch, and so on) in agriculture. The biomass resources are converted to sugar by the saccharification firstly, and then the sugar are converted to ethanol by the fermentation (as shown in route 2).

Biomass → Sugar → Ethanol (Route 2)

The ethanol produced from biomass is called as biomass-derived ethanol or bio-ethanol. Because biomass consumes CO_2 to sustain the growth, using bio-ethanol as a motor fuel can be regarded to reduce CO_2 emission. However, with the increasing application of ethanol as motor fuel, the price of food increases quickly because a part of food is converted to ethanol as motor fuel. Further, the fermentation process is limited in its application only to selected biomass components for ethanol production. Biomass includes various plant components, such as starch, cellulose, hemicellulose, and lignin. While starch and cellulose are biopolymers of glucose, a 6-carbon sugar (hexose), the hemicellulose is primarily pentosans or polymeric pentose (most xylose), and lignin is polymeric phenyl propane. The traditional fermentation technology only can convert 6-carbon sugars from starch and cellulose to ethanol. The current research is focusing on converting 5-carbon sugars from hemicellulose to ethanol. Because lignin (about 30% in woody biomass) can not be converted to a sugar, it is impossible to convert lignin to ethanol by the fermentation process.

Recently, an industry BTL (biomass to liquid) process (as shown in route 3) for convert biomass resources to ethanol has received considerable attention worldwide [7, 8]. This chemical process uses synthesis gas as a platform.

Biomass → Synthesis gas → Ethanol (Route 3)

Synthesis gas (so-called as syngas) is a mixed gas of CO and H_2. The BTL process mainly contains two steps: biomass gasification to produce syngas, and syngas liquefaction to produce ethanol. In the first gasification step, biomass reacts with air (or oxygen) and steam to produce a mixed gas consisting primarily of CO and H_2 through a high-temperature thermochemical process at 973–1173 K. And in the second liquefaction step, the obtained synthesis gas is converted to ethanol in a vapor-solid catalytic system at 473–673 K using solid catalysts.

Synthesis gas can produce various chemicals and fuels (such as hydrocarbons, alcohols, esters, and so on) by using various catalysts. In generally, a H_2/CO ratio of 2 in synthesis gas is needed for the most reactions of synthesizing chemicals and fuels. In the BTL process, the H_2/CO ratio in the synthesis gas obtained from biomass gasification can be adjusted to 2 using the water-gas shift reaction (as shown in formula 1) in order to fit to the following catalytic reactions.

$$CO + H_2O \leftrightarrow CO_2 + H_2 \quad \text{(Formula 1)}$$

Because synthesis gas can be obtained from all kinds of plant components (including lignin) by gasification, the BTL process can convert all kinds of biomass resources to ethanol. Actually, all kinds of organic resources (not only biomass, but also coal, natural gas, asphalt, plastic, and so on) can be converted to synthesis gas by the gasification and/or reforming (with steam, oxygen, or CO_2). The formed synthesis gas can be converted to various fuels and chemicals by using various catalysts. Using the synthesis gas platform, the conversion of biomass to liquid fuels is called as BTL process, the conversion of coal to liquid fuels is called as CTL process, and the conversion of natural gas to liquid fuels is called as GTL process. The liquid fuels including hydrocarbons (such as gasoline, diesel, and so on), alcohols (such as methanol, ethanol, and so on), esters (such as DME, ETBE, and so on). All reactions for converting synthesis gas to fuels or chemicals are catalytic reactions and thus the performances of catalysts determine whether the process can be commercialized or not.

The methanol synthesis (as shown in formula 2) and the Fischer–Tropsch synthesis (as shown in formula 3) became the basis of the C1 chemistry since they had been commercialized in very huge scales.

$$CO + 2\,H_2 \leftrightarrow CH_3OH \qquad \text{(Formula 2)}$$

$$n\,CO + (2n+1)\,H_2 \leftrightarrow C_nH_{2n+2} + n\,H_2O \qquad \text{(Formula 3)}$$

The Cu catalyst has high catalytic performance for the methanol synthesis and the Fe, Co, and Ru catalysts have high catalytic performances for the Fischer–Tropsch synthesis. These highly efficient catalysts are the key factors for the large-scale industrialization of the Fischer–Tropsch (F–T) synthesis and the methanol synthesis. At the present time, almost all worldwide methanol production is obtained from the methanol synthesis process (formula 2), and almost all non-petroleum-derived gasoline and diesel are produced from the F–T synthesis process (formula 3).

The synthesis of ethanol and the synthesis of mixed alcohols from synthesis gas are carried out according to formula 4 and formula 5.

$$2\,CO + 4\,H_2 \leftrightarrow C_2H_5OH + H_2O \qquad \text{(Formula 4)}$$

$$n\,CO + 2n\,H_2 \leftrightarrow C_nH_{2n+1}OH + (n-1)\,H_2O \qquad \text{(Formula 5)}$$

Four types of catalysts have been used in the synthesis of higher alcohols (including ethanol) from synthesis gas: modified F–T synthesis catalysts (containing Ni, Fe, Co, or Ru but not containing Cu, Rh, or Mo) [9, 10], modified Cu-based methanol synthesis catalysts [11–20], Rh-based catalysts [21–32], and MoS-based catalysts [33–43]. Among them, the modified F–T synthesis catalysts almost are not attractive recently because they produce hydrocarbons as main products and the fraction of alcohols in the oxygenated products are quite low. On the other hand, many researches are focusing on Cu-based methanol synthesis catalysts, Rh-based catalysts, and MoS-based catalysts in the synthesis of higher alcohols (including ethanol) from synthesis gas [11–43]. However, the catalytic performances of the reported catalysts are still not high enough to achieve a large-scale commercialization for the ethanol synthesis from syngas. Hence, the development of high efficient catalysts for the synthesis of higher alcohols (including ethanol) from synthesis gas has become an urgent research subject on a worldwide level.

In the present study, we investigated the catalytic abilities of Cu-based, Rh-based, MoS-based for the synthesis of mixed alcohols (including ethanol) in a vapor-solid catalytic system. By adding promoters and adjusting supports, we developed three types of high efficient catalysts for the synthesis of mixed alcohols from sygas.

EXPERIMENTAL

Reaction System

Figure 1 shows the flow chart of the high-pressured fixed-bed flowed reaction system used in this study. A stainless steel tubular reactor (*i.d.*: 1 cm; length: 50 cm) was used for loading catalyst (1–5 g) and a furnace was used for heating the tubular reactor. During the reaction, the feed syngas (containing 30% CO, 60% H_2, 10% N_2) was introduced into the tubular reactor from a high-pressure cylinder and the flow rate was controlled by a mass flow controller. The pressure in the reaction system was controlled by a back-pressure regulator. A cold trap which soaked in a tank of ice water was set behind the back-pressure regulator to collect liquid products.

The products were continuously analyzed by an on-line analysis system containing four GCs during the reaction. H_2, N_2, CO, and CO_2 were analyzed by a TCD-GC equipped with a New Carbon-ST column; light hydrocarbons (C_1–C_4) were analyzed by a FID-GC equipped with a CP-Al_2O_3/KCl capillary column; heavy hydrocarbons (C_{5+}) were analyzed by a FID-GC equipped with a UA-DX capillary column; and oxygenated compounds were analyzed by a FID-GC equipped with a Stablewax capillary column. The factors of various compounds were obtained using a standard mixed gas (with known concentration for each component) from a cylinder.

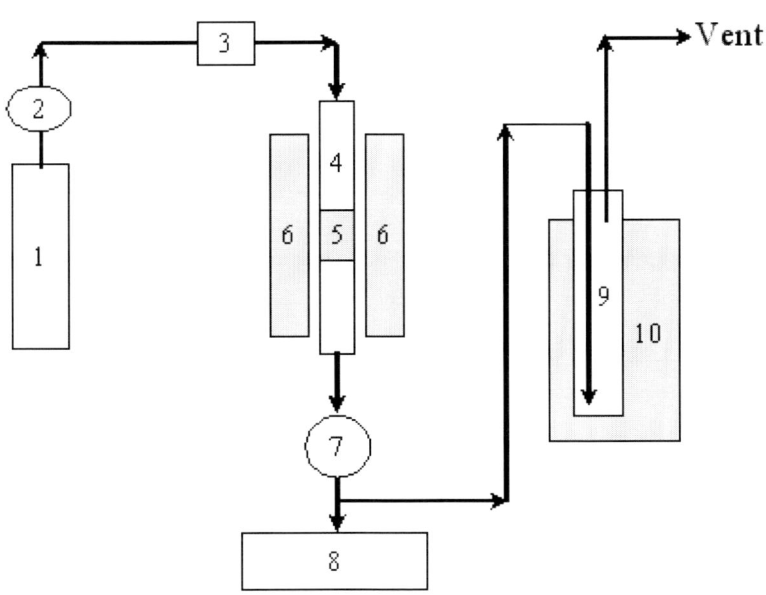

Figure 1. Flow chart of high-pressured fixed-bed flowed reaction system used in this study. (1) Syngas cylinder (30% CO, 60% H_2, 10% N_2), (2) reducing valve, (3) mass flow controller, (4) tube stainless steel reactor, (5) catalyst layers, (6) furnaces, (7) back-pressure regulator, (8) on-line analysis system, (9) cold trap, (10) ice-water tank.

CO conversion was determined by the changes of the percentage concentrations of CO and N_2 (internal standard) in the mixed gases before and after the reaction. Product selectivities were reported in terms of carbon efficiencies which were calculated by using the formula $n_iC_i/\Sigma(n_iC_i)$, where n_i is the carbon atom number of the compound. The analytical results satisfied the carbon monoxide material balance adequately.

Catalyst

Cu/CeO_2 was prepared by a coprecipitation-decomposition method using $NH_3{\cdot}H_2O$ as a precipitant in order to avoid the residues of alkali metals [19]. Some Cu^{2+} ions formed deep blue $[Cu(NH_3)_4]^{2+}$ complexes with NH_3 during the precipitation process, so we designed a boiling step followed the precipitation step to eliminate NH_3 and decompose the complex. In a typical process, $Ce(NO_3)_3{\cdot}6H_2O$ and $Cu(NO_3)_2{\cdot}3H_2O$ were dissolved in distillate water in a beaker. A portion of 28 wt% $NH_3{\cdot}H_2O$ was added dropwise to the aqueous solution of mixed metallic nitrates with strong stirring at room temperature until the pH reached 9.5. Then, the slurry was boiled up for 6 h with stirring till pH = 5.5. Distillate water was added to the beaker to keep the liquid level during the boiling process. The resulting precipitate was then filtered off, washed with distilled water, dried at 373 K for 12 h, and finally calcined at 723 K for 3 h. The Cu loading was 20 wt% in the sample.

Cu/ZnO was prepared using the coprecipitation-decomposition method through an aurichalite precursor [11, 12]. The Cu loading was 20 wt% in the sample.

Cs-Cu/CeO_2 and Cs-Cu/ZnO were prepared by impregnating Cu/CeO_2 and Cu/ZnO in CsOH aqueous solution [11, 12]. After the water was eliminated by heating at 363 K, the solid sample was dried at 373 K for 12 h and calcined at 723 K for 3 h. The Cu loading was 20 wt% and the Cs loading was 1 wt% in the samples.

Cs-Fe-Cu/CeO_2 was prepared by impregnating Cs-Cu/CeO_2 in a $Fe(NO_3)_3$ aqueous solution. After the water was eliminated by heating at 363 K, the solid sample was dried at 373 K for 12 h and calcined at 723 K for 3 h. The Cu loading was 20 wt%, the Cs loading was 1 wt%, and the Fe loading was 3 wt% in the sample.

$Cu/Ce_{1-x}Zr_xO_2$ (x = 0–1) was prepared by the coprecipitation-decomposition method using $NH_3{\cdot}H_2O$ as the precipitant. $Ce(NO_3)_3{\cdot}6H_2O$, $ZrO(NO_3)_2{\cdot}8H_2O$, and $Cu(NO_3)_2{\cdot}3H_2O$ were dissolved in distilled water in a beaker. $NH_3{\cdot}H_2O$ was added dropwise to the aqueous solution of mixed metallic nitrates with strong stirring at room temperature until the pH reached 9.5. The slurry was then boiled for 6 h with stirring until the pH reached 5.5. Distilled water was added to the beaker to maintain the liquid level during the boiling process. The resulting precipitate was then filtered off, washed with distilled water, dried at 373 K for 12 h, and finally calcined at 723 K for 3 h. The Cu loading was 20 wt% in the sample.

Cs-Fe-$Cu/Ce_{1-x}Zr_xO_2$ was prepared by a sequential impregnation method: firstly impregnating $Cu/Ce_{1-x}Zr_xO_2$ in a CsOH aqueous solution and then impregnating Cs-$Cu/Ce_{1-x}Zr_xO_2$ in a $Fe(NO_3)_3$ aqueous solution. After the water was eliminated by heating at 363 K, the solid sample was dried at 373 K for 12 h and calcined at 723 K for 3 h. The Cu loading was 20 wt%, the Cs loading was 1 wt%, and the Fe loading was 3 wt% in the sample.

Rh/SiO_2 was prepared by the impregnation of SiO_2 (JRC-SIO-1, 300 $m^2\ g^{-1}$) in an aqueous solution of $Rh(NO_3)_3$. The sample was then dried at 373 K for 24 h, and finally calcined at 723 K for 3 h. The Rh loading was 2 wt% in the sample.

$Rh/Ce_{1-x}Zr_xO_2$ (x = 0–1) was prepared by a coprecipitation-decomposition method using $NH_3{\cdot}H_2O$ as a precipitant. A solution of 28 wt% $NH_3{\cdot}H_2O$ was added to an aqueous solution of mixed metallic nitrates of $Ce(NO_3)_3$, $ZrO(NO_3)_2$, and $Rh(NO_3)_3$ at room temperature with strong stirring until the pH reached 9.5. The slurry was then boiled for 6 h with stirring until the pH reached 5.5.

A precipitate was obtained after filtering the slurry at room temperature. The resultant precipitate was dried at 373 K for 24 h and calcined at 723 K for 3 h. The Rh loading was 2 wt% in the sample.

Li-Fe-$Rh/Ce_{1-x}Zr_xO_2$ (x = 0–1) catalysts were synthesized by a sequential impregnation method: firstly impregnating $Rh/Ce_{1-x}Zr_xO_2$ in a $Fe(NO_3)_3$ aqueous solution and then impregnating Fe-$Rh/Ce_{1-x}Zr_xO_2$ in a $LiNO_3$ aqueous solution. After the water was eliminated by heating at 363 K, the solid sample was dried at 373 K for 24 h and calcined at 723 K for 3 h. The loading of Rh was 2 wt%, the loading of Fe was 5 wt%, and the loading of Li was 1 wt% in the samples.

$[AlO_4Al_{12}(OH)_{24}(H_2O)_{12}]^{7+}$-pillared montmorillonite (abbreviated as Al_{13}-Mont) was prepared by a ion-exchange method [44–46]. The pillaring solution of $[AlO_4Al_{12}(OH)_{24}(H_2O)_{12}]^{7+}$ was prepared by the dropwise addition of a solution of 0.2 M NaOH (2.4 g NaOH in 300 ml H_2O) to a solution of 0.5 M $Al(NO_3)_3$ (9.38 g $Al(NO_3)_3{\cdot}9H_2O$ in 50 ml H_2O) until a OH/Al molar ratio of 2.4.

After standing at room temperature for 24 h, 5 g Na-type montmorillonite (denoted as Na-Mont, a natural clay mineral with a cation exchange capacity of 100 meq/100 g) was added to the pillaring solution.

The final suspension was stirred at room temperature for 12 h and a solid sample was obtained by filtration. The solid sample was dried at 373 K for 24 h and calcined at 723 K for 3 h.

MoS/SiO_2, MoS/C, and MoS/Al_{13}-Mont were synthesized by a wetness impregnation method [47–49]. The support (SiO_2, C, or Al_{13}-Mont) was added to an aqueous solution of $(NH_4)_6Mo_7O_{24}{\cdot}4H_2O$ to form a slurry. The slurry was stirred at 363 K until formation of a solid sample by evaporating water. The solid sample was dried at 373 K for 24 h and calcined at 723 K for 3 h. The loading of MoO_3 was 15 wt % in each sample. Prior to the reaction, the samples were sulfided in a mixed gas containing 10% H_2S and 90% H_2 (flow rate: 60 mL min^{-1}) at 673K for 10 h.

K-Ni-MoS/Al_{13}-Mont was synthesized by a sequential impregnation method: firstly impregnating MoS/Al_{13}-Mont in a KNO_3 aqueous solution and then impregnating K-MoS/Al_{13}-Mont in a $Ni(NO_3)_2$ aqueous solution. After the water was eliminated by heating at 363 K, the solid sample was dried at 373 K for 12 h and calcined at 723 K for 3 h. The loading of K was 3 wt%, the loading of Ni was 5 wt%, and the loading of MoO_3 was 15 wt% in the sample. Prior to the reaction, the sample was sulfided in a mixed gas containing 10% H_2S and 90% H_2 (flow rate: 60 mL min^{-1}) at 673K for 10 h.

The Cu particle sizes of the Cu supported catalysts were measured by the CO adsorption method by a flow technique at room temperature using a BELCAT-B automatic absorption instrument.

Inductively coupled plasma (ICP) analyses were measured by a Thermo Jarrel Ash IRIS/AP. BET (Brunauer–Emmett–Teller) surface areas of the catalysts were measured by the N_2 adsorption method at 77 K using a Belsorp 28SA automatic absorption instrument.

RESULTS AND DISSUSION

Synthesis of Mixed Alcohols over Modified Cu-Based Catalysts

Table 1 shows the catalytic performances of methanol synthesis catalysts at 573 K. Cu/CeO_2 showed a CO conversion of 23.5%, which was higher than that over the industrial methanol synthesis catalyst Cu/ZnO (17.2%). The selectivity for alcohols over Cu/CeO_2 catalyst was 82.5%, which was also higher than that over Cu/ZnO (78.4%). Methane and carbon dioxide were the main by-products in the synthesis of mixed alcohols over Cu-based catalysts. The mass ratio of C_{2+}OH to MeOH was 0.13 over Cu/CeO_2 and was 0.05 over Cu/ZnO. This indicates that Cu/CeO_2 formed a larger amount of C_{2+}OH higher alcohols than that over Cu/ZnO although both Cu/ZnO and Cu/CeO_2 produced methanol as a dominant product. As a result, as comparison with the industrial methanol synthesis catalyst Cu/ZnO, Cu/CeO_2 is more suitable to be used for the synthesis of higher alcohols by modifying with promoters.

Both Cu^0 species and Cu^+ species are important in the synthesis of methanol over Cu-based catalysts. It has been concluded that the support effect of a Cu-based catalyst in the synthesis of methanol is decided by the amount ratio of Cu^+ to Cu^0 on the support surface [50]. It is likely that the reaction active intermediates are adsorbed on the surface of Cu^0 species, and the role of Cu^+ species is the stabilization of any oxygenates near the Cu^+ sites [50]. Because CeO_2-based compounds have the reducibility and a redox equilibrium ($Ce^{4+} + Cu^0 \leftrightarrow Ce^{3+} + Cu^+$) exists between CeO_2 support and Cu catalyst in Cu/CeO_2, both Cu^0 species and Cu^+ species can be stabilized in the Cu/CeO_2 catalyst [51–54].

As a result, the stabilization of Cu^0 species and Cu^+ species at the same time effectively improved the activity of the Cu/CeO_2 catalyst for the synthesis of methanol from syngas (table 1).

Table 2 shows the catalytic performances of Cs-modified Cu-supported catalysts at 573 K. The mass ratio of C_{2+}OH to MeOH over Cs-Cu/ZnO was 0.53, which was much higher than that over Cu/ZnO (0.05) (table 1).

Moreover, by introducing 1.0 wt% Cs in Cu/CeO_2, the mass ratio of C_{2+}OH to MeOH greatly increased from 0.13 to 0.71 (table 1). On the other hand, the CO conversion over Cs-Cu/ZnO (16.5%) was slightly lower than that over Cu/ZnO (17.2%), and the CO conversion over Cs-Cu/CeO_2 (22.2%) was also slightly lower than that over Cu/CeO_2 (23.5%). As a result, Cs promoter greatly improved the formation of C_{2+} higher alcohols but slightly impressed the catalyst activity over the Cu-based catalysts.

Table 1. Catalytic performance of methanol synthesis catalysts at 573 K[a]

Catalyst [b]	CO Conv. (C%)	Selectivity (C%) Alcohols	Hydrocarbons	CO_2	C_{2+}OH/ MeOH [c]
Cu/ZnO	17.2	78.4	4.6	17.0	0.05
Cu/CeO_2	23.5	82.5	2.6	14.9	0.13

[a] Reaction conditions: $n(H_2)/n(CO)=2$; P=3 MPa; GHSV=2400h^{-1}.

[b] Cu: 20wt%.

[c] Mass ratio of higher alcohols to methanol in product.

Table 2. Catalytic performance of Cs-modified Cu-supported catalysts at 573 K[a]

Catalyst [b]	CO Conv. (C%)	Selectivity (C%)			$C_{2+}OH$/ MeOH [c]
		Alcohols	Hydrocarbons	CO_2	
Cs-Cu/ZnO	16.5	77.9	4.8	17.3	0.53
Cs-Cu/CeO_2	22.2	81.3	2.5	16.2	0.71

[a] Reaction conditions: $n(H_2)/n(CO)=2$; $P=3$ MPa; GHSV=2400h^{-1}.
[b] Cu: 20wt%, Cs: 1wt%.
[c] Mass ratio of higher alcohols to methanol in product.

Scheme 1 outlines the reaction pathway of CO hydrogenation over Cs-Cu-based catalysts. Cs is an excellent promoter to improve the selectivity for higher alcohols over the Cu/ZnO catalyst [11, 12]. At first, methanol is formed on the Cu sites from CO and H_2. Then, ethanol was formed via direct methanol coupling reaction.

The pathway involves necleophilic attack of formaldehyde by adsorbed formyl species, with both species formed preferentially from methanol. Cs^+ is required to stabilize the required formyl species. K^+, Zn^{2+}, and Cu surface atoms are not able to stabilize the intermediates required for methanol-coupling pathways. However, the formed ethanol further reacts with formyl species to form propanol, butanol, and so on. Hence, Cs-Cu-based catalysts produce mixed alcohols from syngas.

Cs-Cu/ZnO is a standard modified methanol catalyst for the synthesis of mixed alcohols from syngas [12]. As shown in table 2, the Cs-Cu/CeO_2 catalyst synthesized in this study showed a higher CO conversion and a higher mass ratio of $C_{2+}OH$ to MeOH than those over the Cs-Cu/ZnO catalyst. Hence, Cs-Cu/CeO_2 is a catalyst more effective than the reported Cs-Cu/ZnO catalyst for the synthesis of higher alcohols from syngas. Further, we tried to improve the catalytic performance of Cs-Cu/CeO_2 for the synthesis of higher alcohols from syngas by adding F-T elements (Fe, Co, and Ni) and adjusting support components.

Table 3 shows the catalytic performances of F-T elements modified Cs-Cu/CeO_2 for the synthesis of mixed alcohols from syngas at 573 K. By introducing 3 wt% Ni in Cs-Cu/CeO_2, the CO conversion increased from 22.2% to 37.6% but the selectivity for total alcohols decreased from 81.3% to 74.1% (due to the formation of methane).

The mass ratios of $C_{2+}OH$ to MeOH increased from 0.71 to 0.85 after introducing 3 wt% Ni in Cs-Cu/CeO_2. On the other hand, Cs-Co-Cu/CeO_2 showed the highest CO conversion (46.6%) but the selectivity for alcohols over Cs-Co-Cu/CeO_2 was low (65.1%) because of the formation of C_{2+} hydrocarbons.

Among various catalysts, Cs-Fe-Cu/CeO_2 showed the highest ratio of $C_{2+}OH$ to methanol (1.03) with a high selectivity for total alcohols (80.5%). As a result, Fe is an effective promoter to increase the formation of $C_{2+}OH$ higher alcohols over Cs-Cu/CeO_2 in the synthesis of mixed alcohols from syngas.

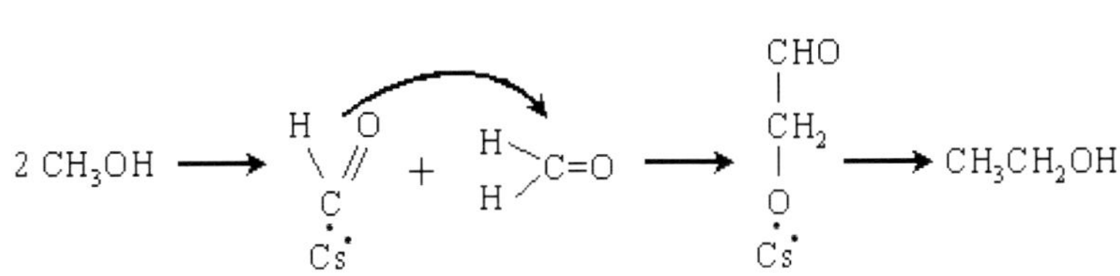

Scheme 1. Reaction pathway of CO hydrogenation over Cs-Cu-based catalysts.

Table 3. Catalytic performances of F-T elements modified Cs-Cu/CeO_2 for the synthesis of mixed alcohols from syngas at 573 K [a]

Catalyst [b]	CO Conv. (C%)	Selectivity (C%) Alcohols	Hydrocarbons	CO_2	$C_{2+}OH$/ MeOH [c]
Cs-Cu/CeO_2	22.2	81.3	2.5	16.2	0.71
Cs-Ni-Cu/CeO_2	37.6	74.1	5.9	20.0	0.85
Cs-Co-Cu/CeO_2	46.6	65.1	9.4	25.5	0.96
Cs-Fe-Cu/CeO_2	40.5	80.5	3.8	15.7	1.03

[a] Reaction conditions: $n(H_2)/n(CO)=2$; P=3 MPa; GHSV=2400h^{-1}.
[b] Cu: 20wt%; Cs: 1wt%; Ni, Co, or Fe (if applicable): 3wt%.
[c] Mass ratio of higher alcohols to methanol in product.

Scheme 2 outlines the reaction pathway of CO hydrogenation over Fe-Cu-based catalysts. Introduction of F-T elements (Fe, Co, and Ni) to Cu-based catalysts is a method to increase the formation of higher alcohols production by increasing the C-C chain [16].The active iron species act as sites for CO dissociation, C–C chain growth, and hydrogenation, while copper sites adsorb CO molecularly. CO* molecule moves to an adsorbed $*CH_x$ group and inserts between metal site and the alky group via surface migration over a short distance between Fe and Cu sites, which is further hydrogenated to form ethanol. The carbon-chain growth of the alkyl group ($*C_nH_z$) is propagated via $*CH_x$ addition. Then the direct hydrogenation of $*C_nH_z$ species leads to the paraffin products. On the other hand, the CO insertion of $*C_nH_z$ species and further hydrogenation result in the production of C_{n+1} higher alcohols [20].

Table 4 shows the effect of Zr amount in Cs-Fe-Cu/$Ce_{1-x}Zr_xO_2$ (x = 0–0.5) catalysts for the synthesis of mixed alcohols at 573 K. Both the CO conversion and the ratio of higher alcohols $C_{2+}OH$ to methanol showed the maximum values at x = 0.2 over Cs-Fe-Cu/$Ce_{1-x}Zr_xO_2$. CeO_2-based compounds have high oxygen storage capacity and reducibility [51, 52]. The addition of ZrO_2 to CeO_2 leads to improvement in oxygen storage capacity and reducibility [53]. This is due to the partial substitution of Ce^{4+} by Zr^{4+} in the lattice of CeO_2, resulting in formation of a solid solution. The small Zr^{4+} ions can be introduced in the Ce^{4+} positions till x = 0.2 in Cu/$Ce_{1-x}Zr_xO_2$ [53]. The rest Zr^{4+} ions formed ZrO_2 particles on the catalyst surface when x was larger than 0.2 in Cu/$Ce_{1-x}Zr_xO_2$. Thus both $Ce_{0.7}Zr_{0.3}O_2$ and $Ce_{0.5}Zr_{0.5}O_2$ are not solid solutions. Because $Ce_{0.8}Zr_{0.2}O_2$ is a solid solution with the largest amount of Zr^{4+} ions, it has the highest oxygen storage capacity and reducibility among $Ce_{1-x}Zr_xO_2$. This is the reason in the chemical aspect for the highest catalytic performance of Cs-Fe-Cu/$Ce_{0.8}Zr_{0.2}O_2$ among Cs-Fe-Cu/$Ce_{1-x}Zr_xO_2$ for the synthesis of mixed alcohols from syngas.

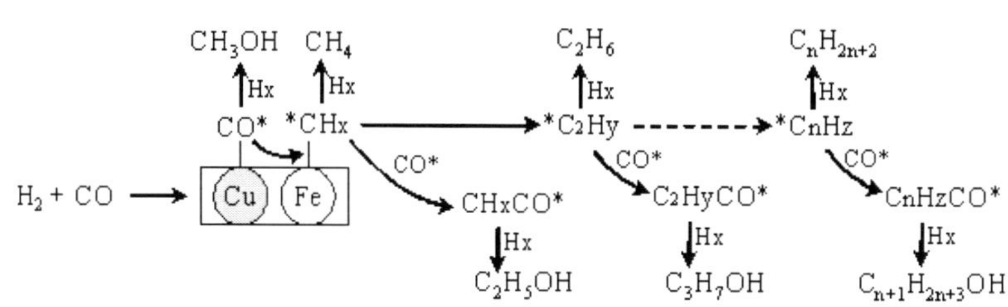

Scheme 2. Reaction pathway of CO hydrogenation over Fe-Cu-based catalysts.

Table 4. Effect of Zr amount in Cs-Fe-Cu/$Ce_{1-x}Zr_xO_2$ (x = 0–0.5) catalysts for the synthesis of mixed alcohols at 573 K[a]

Catalyst [b]	CO Conv. (C%)	Selectivity (C%) Alcohols	Hydrocarbons	CO_2	$C_{2+}OH$/ MeOH [c]
Cs-Fe-Cu/CeO_2	40.5	80.5	3.8	15.7	1.03
Cs-Fe-Cu/$Ce_{0.9}Zr_{0.1}O_2$	43.2	80.9	4.5	14.6	1.07
Cs-Fe-Cu/$Ce_{0.8}Zr_{0.2}O_2$	47.3	81.0	4.8	14.2	1.12
Cs-Fe-Cu/$Ce_{0.7}Zr_{0.3}O_2$	45.6	76.5	8.1	15.4	1.06
Cs-Fe-Cu/$Ce_{0.5}Zr_{0.5}O_2$	44.8	72.3	9.8	17.9	0.98

[a] Reaction conditions: $n(H_2)/n(CO)=2$; P=3 MPa; GHSV=2400h^{-1}.
[b] Cu: 20wt%; Cs: 1wt%; Fe: 3wt%.
[c] Mass ratio of higher alcohols to methanol in product.

Figure 2 shows BET surface areas and Cu particle sizes of the Cs-Fe-Cu/$Ce_{1-x}Zr_xO_2$ (x = 0–0.5) catalysts after reduction in H_2 at 573 K for 1 h. Doping Zr^{4+} ions into CeO_2 lattices decreased the crystallization degree. The decrease of crystallization degree increased the BET surface area of the $Ce_{1-x}Zr_xO_2$ support [53]. As shown in figure 2, Cs-Fe-Cu/$Ce_{0.8}Zr_{0.2}O_2$ showed a BET surface area higher than that of CsCu/CeO_2 due to doping Zr^{4+} ions in CeO_2 lattices. Because $Ce_{0.7}Zr_{0.3}O_2$ and $Ce_{0.5}Zr_{0.5}O_2$ are physical mixtures but are not solid solutions, the BET surface areas of Cs-Fe-Cu/$Ce_{0.7}Zr_{0.3}O_2$ and Cs-Fe-Cu/$Ce_{0.5}Zr_{0.5}O_2$ were lower than that of Cs-Fe-Cu/$Ce_{0.8}Zr_{0.2}O_2$. The actual Cu loadings, which were measured by ICP analysis, were same as the designed Cu loadings in various catalysts. The Cu particles' size was calculated by the actual Cu loading and the amount of adsorbed CO molecules for each catalyst. As shown in figure 2, the Cu particles' size showed the minimum value at x = 0.2 in Cs-Fe-Cu/$Ce_{1-x}Zr_xO_2$. Cs-Fe-Cu/$Ce_{0.8}Zr_{0.2}O_2$ had the largest BET surface and the smallest Cu particles among various Cs-Fe-Cu/$Ce_{1-x}Zr_xO_2$. This is the reason in the physical aspect for the highest catalytic performance of Cs-Fe-Cu/$Ce_{0.8}Zr_{0.2}O_2$ among Cs-Fe-Cu/$Ce_{1-x}Zr_xO_2$ for the synthesis of mixed alcohols from syngas.

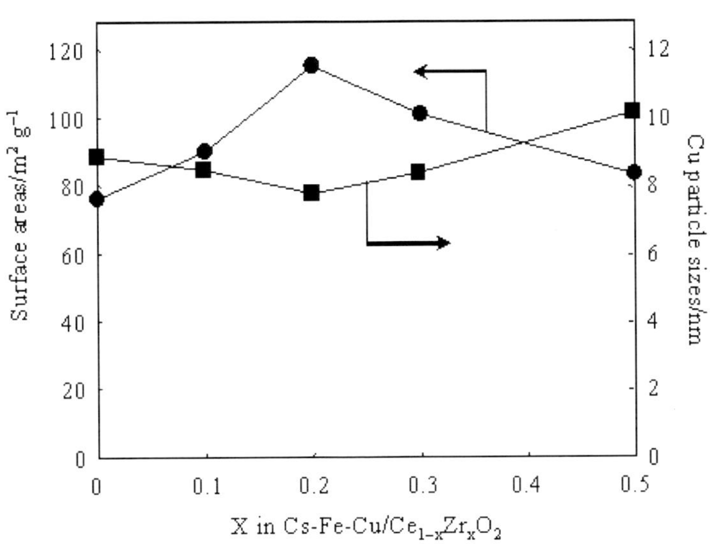

Figure 2. BET surface areas and Cu particle sizes in the Cs-Fe-Cu/$Ce_{1-x}Zr_xO_2$ catalysts after reduction at 573 K.

Table 5 shows the catalytic performance of Cs-Fe-Cu/$Ce_{0.8}Zr_{0.2}O_2$ for the synthesis of mixed catalysts at 523 and 573 K. Reaction at a low temperature of 523 K, the CO conversion was low (12.9%) but the selectivity for total alcohols was high (89.5%). On the other hand, reaction at a high temperature of 573 K formed a relatively large amount of CO_2 (14.2%), which caused a decrease of selectivity for total alcohols. However, the ratio of $C_{2+}OH$ higher alcohols to methanol at 523 K (0.85) was lower than that obtained at 573 K (1.12). As a result, a lower reaction temperature is favorable for the production of methanol and a higher reaction temperature is favorable for the production of higher alcohols.

Figure 3 shows the mass fraction of each alcohol in the total alcohols formed over Cs-Fe-Cu/$Ce_{0.8}Zr_{0.2}O_2$ in the synthesis of mixed alcohols from syngas at 523 and 573 K. When the reaction was carried out at a low temperature of 523 K, the mass fraction of each alcohol product was in an order of $CH_3OH > C_2H_5OH > C_3H_7OH > C_4H_9OH > C_5H_{11}OH > C_6H_{13}OH$. However, when the reaction was carried out at a high temperature of 573 K, C_4OH (almost isobutanol) became the main higher alcohol products in the synthesis of mixed alcohols over Cs-Fe-Cu/$Ce_{0.8}Zr_{0.2}O_2$.

Table 5. Catalytic performance of Cs-Fe-Cu/$Ce_{0.8}Zr_{0.2}O_2$ for the synthesis of mixed catalysts at 523 and 573 K[a]

Temperature (K)	CO Conv. (C%)	Selectivity (C%) Alcohols	Hydrocarbons	CO_2	$C_{2+}OH$/MeOH [b]
523	12.9	89.5	4.6	5.9	0.85
573	47.3	81.0	4.8	14.2	1.12

[a] Reaction conditions: $n(H_2)/n(CO)=2$; P=3 MPa; GHSV=2400h^{-1}; Cu: 20wt%; Cs: 1wt%; Fe: 3wt%.
[b] Mass ratio of higher alcohols to methanol in product.

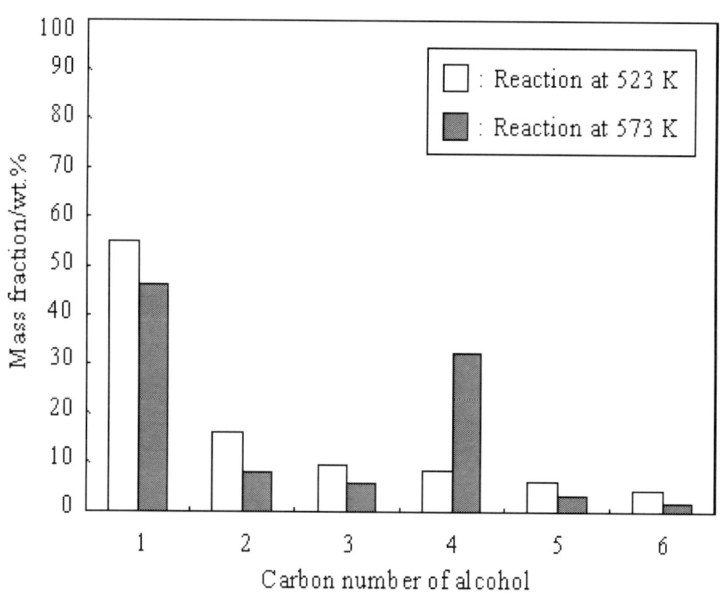

Figure 3. Mass fraction of each alcohol in the total alcohols formed over Cs-Fe-Cu/$Ce_{0.8}Zr_{0.2}O_2$ in the synthesis of mixed alcohols from syngas at 523 and 573 K.

The synthesis of mixed alcohols over modified Cu-based catalysts obeys a CO insertion mechanism (as shown in schemes 1, 2). According to the CO insertion mechanism, the amount of each alcohol in the total alcohols should obey an Anderson-Schultz-Flory (A–S–F) distribution during the reaction. The A–S–F distribution can be expressed as equation 1:

$$W_n/n = (1-\alpha)^2\alpha^{n-1} \quad \text{(Equation 1)}$$

where W_n is the weight fraction of an alcohol (containing n carbon atoms) in the total alcohols in the product. The parameter α is defined as the chain growth probability that a molecule continues reacting to form a longer chain. The parameter α is larger than zero but is lower than one because a part of intermediates desorbed from the catalyst surface in each step of the growth of carbon chain. The amount of each alcohol in the product must decrease with increasing of carbon chain according to A–S–F distribution.

Several factors determine that isobutanol is the second most abundant alcohol while methanol is the dominant alcohol in the synthesis of mixed alcohols over modified Cu-based catalysts at high reaction temperatures. Firstly, the standard heat of enthalpy and the Gibbs free energy of isobutanol are the lowest among C_1–C_4 alcohols [17, 18]. Isobutanol is easy to be formed as comparing with ethanol and propanol from the view point of thermodynamics. Secondly, in each C–C bond forming step ($C_1 \rightarrow C_2$, $C_2 \rightarrow C_3$, $C_3 \rightarrow C_4$), the $C_1 \rightarrow C_2$ step is the slowest one and is the rate-determining step in forming the higher alcohols from syngas [14]. The subsequent $C_2 \rightarrow C_3$ and $C_3 \rightarrow C_4$ steps proceed at faster rates and this leads to a depletion of ethanol in the reaction mixture. In the meantime, branched alcohols are terminal products and this leads to an enrichment of isobutanol in the product mixture. Thirdly, the precursor of ethanol can form butanol by an aldol-type condensation reaction over the basic catalyst surface. Because both Cs and CeO_2 are strong basic sites, Cs-Fe-Cu/$Ce_{0.8}Zr_{0.2}O_2$ showed a higher fraction of isobutanol in the synthesis of mixed alcohols. Fourthly, The CeO_2-based compounds (without supporting Cu metal) are active catalysts for the synthesis of isobutane from syngas (so-called as isosynthesis) due to reducibility and oxygen-storage capacity [55, 56]. Isobutane is a thermodynamically stable hydrocarbon and isobutanol is a thermodynamically stable alcohol.

All modified Cu-based catalysts produce isobutanol as the second most abundant alcohol next to methanol in the synthesis of mixed alcohols from syngas [11–20]. The fraction of isobutanol in total alcohols is important for evaluating a modified Cu-based catalyst in the synthesis of mixed alcohols. The formed isobutanol in the reaction can be used for synthesizing MTBE by reacting with the formed methanol [17, 18]. MTBE has been used as an additive for gasoline in the world.

$$CH_3OH + i\text{-}C_4H_9OH \leftrightarrow CH_3OC_4H_9 + H_2O \quad \text{(Formula 6)}$$

On the whole, our Cs-Fe-Cu/$Ce_{0.8}Zr_{0.2}O_2$ catalyst showed a higher catalytic performance than those of the reported Cu-based catalysts. Because the selectivity for CO_2 and hydrocarbons (almost methane) were low, the mixed alcohols synthesis using the Cs-Fe-Cu/$Ce_{0.8}Zr_{0.2}O_2$ catalyst can be industrialized in the current equipment of methanol synthesis process. However, the process produces methanol and isobutanol as the main products, and ethanol is obtained just as a by-product.

Synthesis of Ethanol over Modified Rh-Based Catalysts

Table 6 shows the reaction results of ethanol synthesis from syngas over various catalysts with 2 wt% Rh loading at 548 K. Rh/CeO_2 showed a CO conversion of 23.7%, which was much higher than those over Rh/SiO_2 (10.1%), Rh/MgO (10.8%), and Rh/ZrO_2 (18.2%). Further, Rh/$Ce_{0.8}Zr_{0.2}O_2$ showed a higher conversion (27.3%) than that over Rh/CeO_2 (23.7%) and Rh/ZrO_2 (18.2%). CeO_2-supported metal catalysts possess a strong interaction between support and metal (SISM), which give these catalysts high catalytic activity for some reactions [51–53, 57–61]. The SISM effect is important for improving catalytic activity in the synthesis of ethanol from syngas over Rh-based catalysts [27]. The SISM effect contains several factors and the reducibility of support is an important factor for the Rh-based catalysts [26]. We think that the reducibility of CeO_2 support gave Rh/CeO_2 a higher CO conversion for the synthesis of ethanol from syngas. It has been reported that (Rh_x^0–Rh_y^+)–O–M is the active site for the formation of C_2-oxygenates [28]. Thus both Rh^0 species and Rh^+ species are important in the synthesis of ethanol from syngas [28]. Because a redox equilibrium (Ce^{4+} + $Rh^0 \leftrightarrow Ce^{3+}$ + Rh^+) exists between CeO_2 support and Rh particles, both Rh^0 species and Rh^+ species can be stabilized in Rh/CeO_2. Moreover, the introduction of Zr^{4+} ions into Rh/CeO_2 improves the reductibility of CeO_2 support. The Rh^+ active species became more stable in the Rh/$Ce_{0.8}Zr_{0.2}O_2$ catalyst due to the high reductibility of $Ce_{0.8}Zr_{0.2}O_2$ support, which caused an increase of the SISM effect in the catalysts [31, 53]. As a result, Rh/$Ce_{0.8}Zr_{0.2}O_2$ showed a high CO conversion in the synthesis of ethanol from syngas because the $Ce_{0.8}Zr_{0.2}O_2$ support had high reductibility.

As shown in table 6, Rh/SiO_2 formed C_2-oxygenates (ethanol, acetaldehyde, and AcOEt) as the main oxygenated products and the selectivity to methanol was low. Rh/ZrO_2 also formed the mixture of C_2-oxygenates as the main oxygenated products. The selectivities for esters and ethers over Rh/ZrO_2 were higher than those over Rh/SiO_2 due to the acidity of ZrO_2 support. In contrast, Rh catalysts supported on basic supports MgO and CeO_2 formed mixed alcohols as the main oxygenated products and the selectivities for acetaldehyde and AcOEt were low. Hence, Rh catalysts supported on neutral or acidic supports formed the mixture of C_2-oxygenates and Rh catalysts supported on basic supports formed mixed alcohols from syngas. The amount of hydrocarbons formed over each Rh-based catalyst was in an order of Rh/ZrO_2 > Rh/SiO_2 > Rh/MgO > Rh/CeO_2.

Table 6. Reaction results of ethanol synthesis from synthesis gas over various catalysts with 2 wt% Rh at 548 K[a]

Catalyst	Conv. (C%)	Oxygenate selectivity /% MeOh	EtOH	CH_3CHO	Other [b]	HC select./% CH_4	C_{2+}	Co_2 select./%
Rh/SiO_2	10.1	3.3	16.2	13.3	9.6	42.9	12.5	2.2
Rh/ZrO_2	18.2	1.9	15.7	6.3	16.6	48.2	9.8	2.5
Rh/MgO	10.8	34.7	20.1	1.7	3.8	36.1	1.3	1.9
Rh/CeO_2	23.7	15.3	25.4	0.7	7.7	34.2	1.7	15.3
Rh/$Ce_{0.8}Zr_{0.2}O_2$	27.3	7.9	35.2	1.4	7.6	35.7	2.1	10.1

[a] P=2.4 MPa; GHSV=2400 h^{-1}; CO: 30%; H_2:60%; N_2: 10%.

[b] Other oxygenates: C_3H_7OH, $CH_3COOC_2H_5$, CH_3COOCH_3, $HCOOCH_3$, $HCOOC_2H_5$, CH_3OCH_3, $CH_3OC_2H_5$; $C_2H_5OC_2H_5$, and CH_3COOH.

The Rh particles supported basic supports formed a small amount of hydrocarbons in the synthesis of ethanol from syngas. The Rh-based catalysts mainly formed C_2-oxygenates and the fraction of each alcohol in total alcohols did not obey the A–S–F distribution. Hence, the reaction route of Rh-based catalysts is completely different from that of Cu-based catalysts in the synthesis of alcohols from syngas. The Rh-based catalysts directly formed C_2-oxygenated compounds from syngas and did not react via a chain growth mechanism like that in the Cu-based catalysts.

Scheme 3 outlines the reaction route for the formation of C_2-oxygenates from syngas over the Rh-based catalysts. During the reaction, CO and H_2 molecules were adsorbed on Rh sites to form CO_{ad} and H_{ad} species, and then one CO_{ad} reacted with two H_{ad} to form a $CHOH_{ad}$ intermediate on the catalyst surface [21, 59]. The $CHOH_{ad}$ intermediate could form a CH_3OH molecule by hydrogenation or form a $(CH_2)_{ad}$ intermediate by hydrogenation and dehydration [59].

Further, the $(CH_2)_{ad}$ intermediates could form CH_4 by hydrogenation or form C_2-oxygenated compounds by CO insertion [24, 28]. The acidic supports promoted the dehydration process of forming $(CH_2)_{ad}$ intermediates from $CHOH_{ad}$ intermediates, which increased the concentration of $(CH_2)_{ad}$ intermediates in the catalytic system, and thus increased the selectivities for C_2-oxygenated compounds and CH_4. In contrast, the basic supports inhibited the dehydration process from $CHOH_{ad}$ intermediates to $(CH_2)_{ad}$ intermediates, which increased the concentration of $CHOH_{ad}$ intermediates in the catalytic system, and thus increased the selectivity for methanol [31].

Because $Rh/Ce_{0.8}Zr_{0.2}O_2$ had strong base sites on the surface like Rh/CeO_2, $Rh/Ce_{0.8}Zr_{0.2}O_2$ formed mixed alcohols as main oxygenated products like Rh/CeO_2 (as shown in table 6). However, the selectivity for ethanol over $Rh/Ce_{0.8}Zr_{0.2}O_2$ (35.2%) was higher than that over Rh/CeO_2 (25.4%). $Rh/Ce_{0.8}Zr_{0.2}O_2$ had acid sites by introducing Zr^{4+} ions into the CeO_2 lattices.

As shown in scheme 3, the acid sites could promote the dehydration process of $CHOH_{ad}$ intermediates and increase the concentration of $(CH_2)_{ad}$ intermediates in the catalytic system. Because $CHOH_{ad}$ is a precursor of methanol and $(CH_2)_{ad}$ is a precursor of C_2-oxygenates (by CO insertion), $Rh/Ce_{0.8}Zr_{0.2}O_2$ showed a higher selectivity for ethanol than that over Rh/CeO_2. As a result, $Rh/Ce_{0.8}Zr_{0.2}O_2$ showed a high selectivity for ethanol in the synthesis of ethanol from syngas because the $Ce_{0.8}Zr_{0.2}O_2$ support contains both basic sites and acidic sites on the surface.

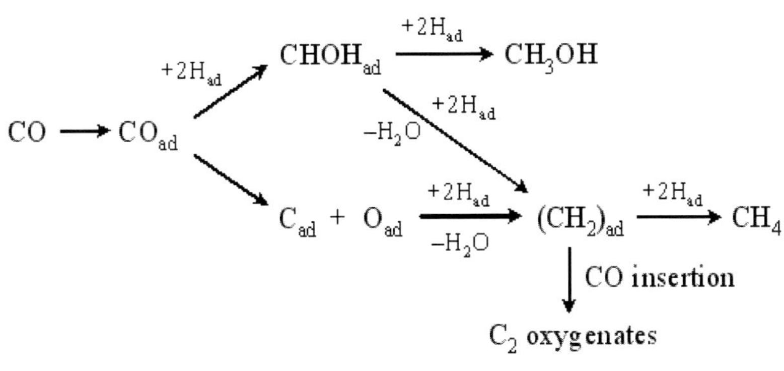

Scheme 3. Reaction pathway in the formation of C_2 oxygenates from syngas over Rh-based catalysts.

Figure 4 shows the effect of x in $Rh/Ce_{1-x}Zr_xO_2$ with 2 wt% Rh loading for the synthesis of ethanol at 548 K. Both the CO conversion and the selectivity for ethanol showed the maximum values at x = 0.2 in $Rh/Ce_{1-x}Zr_xO_2$. Because the small-sized Zr^{4+} ions can be introduced into the Ce^{4+} positions till x = 0.2 in $Rh/Ce_{1-x}Zr_xO_2$, $Ce_{0.8}Zr_{0.2}O_2$ is a solid solution but both $Ce_{0.7}Zr_{0.3}O_2$ and $Ce_{0.5}Zr_{0.5}O_2$ are not solid solutions [53]. The exceed Zr^{4+} ions in $Ce_{0.7}Zr_{0.3}O_2$ and $Ce_{0.5}Zr_{0.5}O_2$ formed ZrO_2 particles on the catalyst surface, which caused a decrease of the CO conversion in the synthesis of ethanol. Moreover, the acidic ZrO_2 particles on the catalyst surface at x > 0.2 in $Rh/Ce_{1-x}Zr_xO_2$ greatly increased the amounts of hydrocarbons and acetaldehyde in the products and decreased the selectivity for ethanol. Therefore, the acid sites are important to form C_{2+} oxygenates from syngas but excess acid sites decrease the selectivity for ethanol due to the formation of hydrocarbons and acetaldehyde. As a result, $Rh/Ce_{0.8}Zr_{0.2}O_2$ showed the highest catalytic performance among various $Rh/Ce_{1-x}Zr_xO_2$ catalysts owing to strong reducibility and properly acidic-basic ability.

Table 7 lists the effect of promoter for the $Rh/Ce_{0.8}Zr_{0.2}O_2$ catalyst in the synthesis of ethanol from syngas at 548 K. In general, Fe catalyst has excellent CO dissociation activity and hence serves as a catalyst for producing linear hydrocarbons (F–T synthesis), while Rh catalyst has excellent CO insertion activity and hence serves as a catalyst for producing C_2 oxygenates. By introducing 5 wt% Fe to $Rh/Ce_{0.8}Zr_{0.2}O_2$, the CO conversion increased from 27.3% to 32.1% and the selectivity for ethanol increased from 35.2% to 42.7%. Therefore, Fe is an effective promoter for the CO hydrogenation to ethanol over supported Rh catalysts.

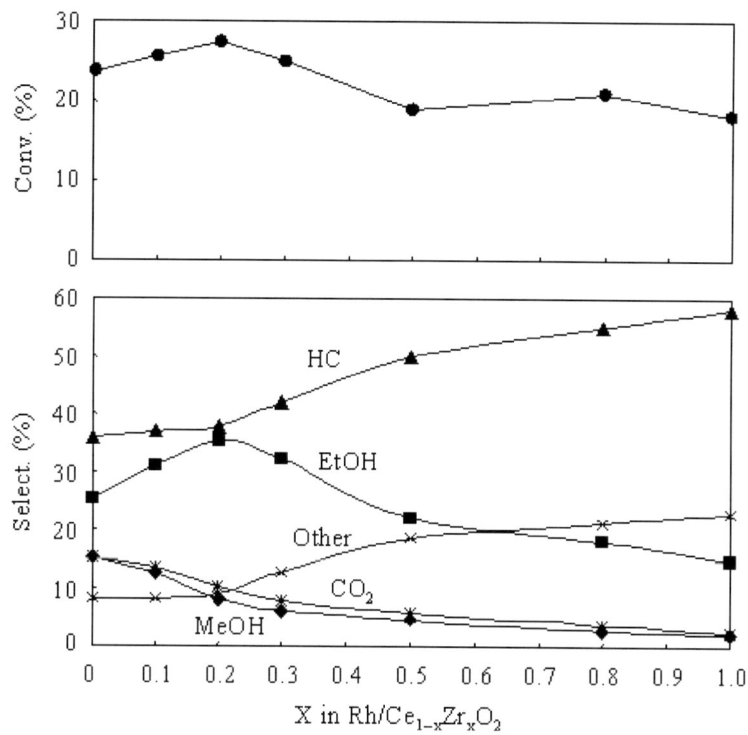

Figure 4. Effect of x in $Rh/Ce_{1-x}Zr_xO_2$ with 2 wt% Rh loading for the synthesis of ethanol at 548 K. (●): CO conversion; (▲): selectivity to hydrocarbons; (■): selectivity to ethanol; (♦): selectivity to methanol; (×): selectivity to other oxygenates; (*): selectivity to CO_2. P = 2.4 MPa; GHSV = 2400 h^{-1}; CO: 30%; H_2: 60%; N_2: 10%.

Table 7. Effect of promoter for the $Rh/Ce_{0.8}Zr_{0.2}O_2$ catalyst in the synthesis of ethanol form syngas at 548 K[a]

Catalyst [b]	Conv./ %	Oxygenate selectivity/%				HC Select./%		CO_2 select./%
		MeoH	EtOH	CH_3CHO	Other [c]	CH_4	C_{2+}	
$Eh/Ce_{0.8}Zr_{0.2}O_2$	27.3	7.9	35.2	1.4	7.6	35.7	2.1	10.1
$Fe\text{-}Rh/Ce_{0.8}Zr_{0.2}O_2$	32.1	4.8	42.7	2.2	6.8	28.3	6.3	8.9
$Li\text{-}Fe\text{-}Rh/Ce_{0.8}Zr_{0.2}O_2$	30.2	5.1	47.4	1.8	6.5	26.6	4.2	8.4

[a] P=2.4 MPa; GSHV=2400 h^{-1}; CO: 30%; H_2: 60%; N_2: 10%.

[b] Rh loading: 2wt%; Fe loading: 5 wt%; Li loading: 1wt%.

[c] Other oxygenates: C_3H_7OH, $CH_3COOC_2H_5$, CH_3COOCH_3, $HCOOCH_3$, $HCOOC_2H_5$, CH_3OCH_3, $CH_3OC_2H_5$; $C_2H_5OC_2H_5$, and CH_3COOH.

Fe improved the step of CO dissociation and thus increased the CO conversion over the $Fe\text{-}Rh/Ce_{0.8}Zr_{0.2}O_2$ catalyst. CH_x (x = 2 or 3) is a common intermediate for forming methane, C_{2+} hydrocarbons, and C_2 oxygenates. Introduction of Fe in $Rh/Ce_{0.8}Zr_{0.2}O_2$ greatly suppressed the formation of methane but the amount of C_{2+} hydrocarbons increased. The selectivity for total hydrocarbons over $Fe\text{-}Rh/Ce_{0.8}Zr_{0.2}O_2$ was lower than that over $Rh/Ce_{0.8}Zr_{0.2}O_2$.

Moreover, CeO_2-supported metal catalysts possess a strong interaction between support and metal (SISM) [57, 58]. The SISM effect gives supported metal a kind of positive charge [57, 58]. Because Fe^0 species is an active species for the F–T synthesis, the SISM effect decreased Fe activity for the F–T synthesis. On the contrary, the SISM effect improved Rh activity for the ethanol synthesis because Rh^+ is a species more active than Rh^0 for the CO insertion.

As shown in table 7, by introducing 1 wt% Li in $Fe\text{-}Rh/Ce_{0.8}Zr_{0.2}O_2$, the CO conversion decreased from 32.1% to 30.2% but the selectivity for ethanol increased from 42.7% to 47.4% (due to inhibition of the selectivity for hydrocarbons). $(Rh_x^0\text{–}Rh_y^+)$–O–M species has been reported as an active site for the formation of C_2 oxygenates [28]. Thus both Rh^0 species and Rh^+ species are important in the synthesis of ethanol from syngas [24]. Rh^0 species are more active for CO absorption and Rh^+ species are more active for CO insertion [24]. Li has an ability to make Rh less reducible, which caused an increase of Rh^+ species and a decrease of Rh^0 species in the Li-modified Rh catalysts. The CH_x (x = 2 or 3) intermediates form methane by hydrogenation and form C_2 oxygenates by CO insertion. Li-modified Rh catalysts showed a high selectivity for ethanol due to the formation of a large amount of Rh^+ species. The CO conversion slightly decreased by adding Li was due to the decrease of Rh^0 species in the catalysts. As a result, Li could be used as unique secondary additives for $Fe\text{-}Rh/Ce_{0.8}Zr_{0.2}O_2$ in the synthesis of ethanol from syngas.

Figure 5 shows the influence of reaction pressure on CO conversion and ethanol selectivity in the synthesis of ethanol from syngas over $Li\text{-}Fe\text{-}Rh/Ce_{0.8}Zr_{0.2}O_2$ at 548 K. Because the residence time of the feed reaction gas pass through the catalyst layer increased with increasing reaction pressure, the CO conversion almost proportionally increased with increasing reaction pressure from 0.1 to 3.6 MPa over $Li\text{-}Fe\text{-}Rh/Ce_{0.8}Zr_{0.2}O_2$. The selectivity for ethanol greatly increased with increasing pressure from 0.1 to 1.2 MPa, and thereafter almost leveled off when the pressure increased from 1.2 to 3.6 MPa.

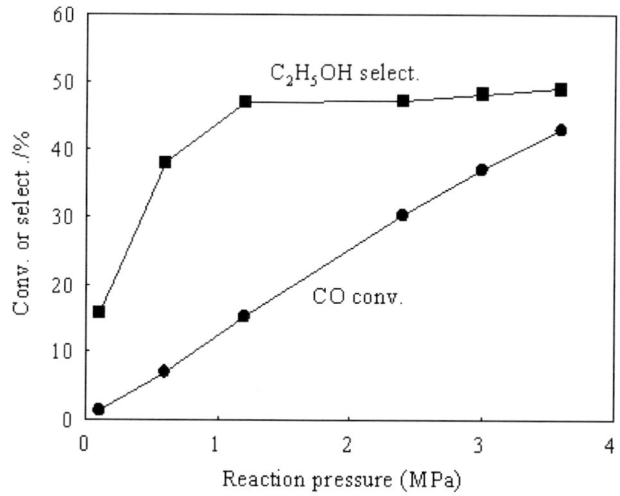

Figure 5. Influence of reaction pressure on CO conversion and ethanol selectivity in the synthesis of ethanol from syngas over Li-Fe-Rh/$Ce_{0.8}Zr_{0.2}O_2$ at 548 K.

In general, the selectivity is the most important factor for evaluating catalytic performance because the conversion can be improved by increasing the amount of catalyst. Therefore, Rh-based catalysts can be used in the synthesis of ethanol from syngas under a relatively low reaction pressure (1–2 MPa). Because biomass and coal have sulfur components, the syngas obtained from the gasification of biomass and coal usually contains sulfur compounds. Here, we added 10 ppm H_2S in the feed syngas to check the influence of sulfur on the Li-Fe-Rh/$Ce_{0.8}Zr_{0.2}O_2$ catalyst in the synthesis of ethanol from syngas.

Figure 6 shows the time sequences in the synthesis of ethanol over Li-Fe-Rh/$Ce_{0.8}Zr_{0.2}O_2$ at 548 K without H_2S or containing 10 ppm H_2S. Using a clear feed gas without any sulfur compounds, Li-Fe-Rh/$Ce_{0.8}Zr_{0.2}O_2$ showed an initial CO conversion of 30.2% and an initial ethanol selectivity of 47.4%. Because both the CO conversion and the ethanol selectivity did not decrease after reaction for 72 hours, Li-Fe-Rh/$Ce_{0.8}Zr_{0.2}O_2$ showed a high catalytic stability in the synthesis of ethanol from clean syngas. In contrast, when 10 ppm H_2S was added in the feed gas, the CO conversion decreased to 20.8% and the ethanol selectivity decreased to 41.1% after reaction for 72 hours. Because H_2S rapidly reacted with Rh to form an inactive RhS phase on the catalyst surface, the speed of deactivation over Li-Fe-Rh/$Ce_{0.8}Zr_{0.2}O_2$ was relatively fast in the first 10 hours and thereafter became slow in the synthesis of ethanol from sulfur-containing syngas.

On the whole, our Li-Fe-Rh/$Ce_{0.8}Zr_{0.2}O_2$ catalysts showed a higher catalytic performance than those of the reported Rh-based catalysts. Li-Fe-Rh/$Ce_{0.8}Zr_{0.2}O_2$ selectively produced ethanol and the amount of formed ethanol was over nine times more than the amount of formed methanol (table 7). The selectivity for by-products hydrocarbons (almost methane) was about 30% over Li-Fe-Rh/$Ce_{0.8}Zr_{0.2}O_2$ although this value was lower than those of the reported catalysts. Li-Fe-Rh/$Ce_{0.8}Zr_{0.2}O_2$ can produce ethanol from syngas even at a low pressure (1–2 MPa), which greatly decreases the equipment cost of the commercial process. We think a large-scale commercial process of ethanol synthesis using the Rh-based catalysts is available if the selectivity for hydrocarbons can be further reduced to about 20%. Because sulfur is a poison for the expensive Rh, it is necessary to eliminate sulfur compounds by a desulfurization process before the sulfur-containing gas has been fed to the Rh-based catalyst.

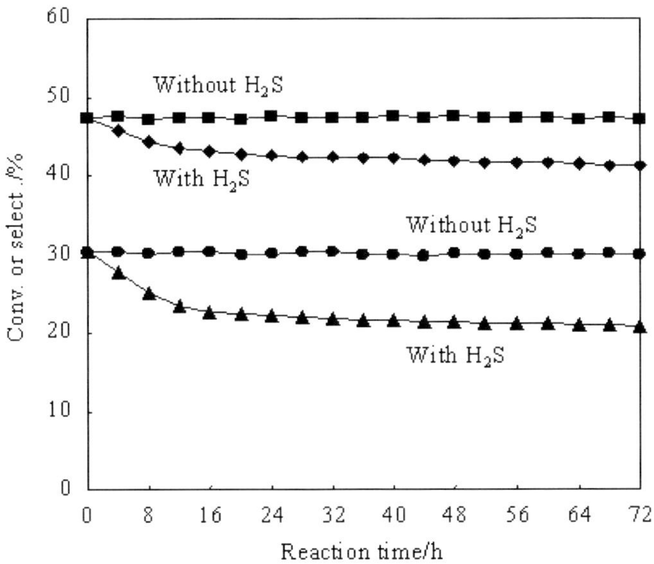

Figure 6. Time sequences in the synthesis of ethanol over Li-Fe-Rh/$Ce_{0.8}Zr_{0.2}O_2$ catalyst at 548 K without H_2S or containing 10 ppm H_2S (●: CO conversion without H_2S; ▲: CO conversion with H_2S; ■: C_2H_5OH selectivity without H_2S; ♦: C_2H_5OH selectivity with H_2S).

Synthesis of Mixed Alcohols over Modified Mos-Based Catalysts

Table 8 shows the catalytic performances of MoS catalysts supported on various supports for the synthesis of mixed alcohols from syngas at 543 K. SiO_2 is the most popular catalyst support and MoS/SiO_2 showed a CO conversion of 15.7% and a selectivity for total alcohols of 23.0% at 543 K. MoS/C showed a relatively high catalytic performance (CO conversion: 24.5%, selectivity for total alcohols: 30.5%) because active carbon is an effective support for the MoS catalyst in the synthesis of mixed alcohols [33]. Montmorillonite (Mont) is a kind of clays which has been reported as an effective support for MoS catalyst in the synthesis of mixed alcohols [42, 43]. MoS/Mont showed a CO conversion of 26.7% and a selectivity for total alcohols of 33.4% over at 543 K. We used MoS/Al_{13}-Mont as a catalyst for the synthesis of mixed alcohols and its catalytic performance (CO conversion: 34.6%, selectivity for total alcohols: 33.5%) was higher than those of the three reported catalysts.

Table 8. Catalytic performances of MoS catalysts supported on various supports for the synthesis of mixed alcohols from syngas at 543 K[a]

Catalyst [b]	CO Conv. (C%)	Selectivity (C%) Alcohols	Hydrocarbons	CO_2	$C_{2+}OH$/ MeOH [c]
MoS/SiO_2	15.7	23.0	44.9	32.1	0.83
MoS/C	24.5	30.5	40.7	28.8	0.75
MoS/Mont	26.7	33.4	42.8	23.8	1.12
Mos/Al_{13}-Mont	34.6	33.5	43.6	22.9	1.16

[a] Reaction condition: $n(H_2)/n(CO)=2$; P=6 MPa; GHSV=2400h^{-1}.

[b] Mo: 15wt%.

[c] Mass ratio of higher alcohols to methanol in product.

Montmorillonite (abbreviated as Mont) is a kind of clay mineral having a two-dimensional layered structure [62, 63]. The layers of Mont have positive charge deficiency and the charge deficiency is balanced by hydrated interlayer cations such as Na^+, K^+, and Ca^{2+}. These small metal cations in the interlayer region of Mont can be replaced by some large cations such as cations of organic compounds, organometallic compounds, and inorganic complexes. Large inorganic metallic polyoxocations (such as $[AlO_4Al_{12}(OH)_{24}(H_2O)_{12}]^{7+}$, abbreviated as Al_{13}) are very attractive as the interlayer cations for Mont. After calcination to eliminate the water between the layers, polyoxocations form pillars to support the silica-alumina layers, and thus the polyoxocation-pillared Monts have been converted into a kind of rigid porous material [64, 65]. Pillared clay Al_{13}-Mont has been used as an effective support for some catalytic reaction [44–46, 62–65].

Table 9 shows the effect of alkali metals on MoS/Al_{13}-Mont for the synthesis of mixed alcohols at 543 K. Alkali metals (Li, K, and Cs) are usually used as promoters on MoS-based catalysts for the synthesis of mixed alcohols [33, 34]. As shown in table 9, the ratio of higher alcohols to methanol slightly increased but both the CO conversion and the selectivity for alcohols decreased by adding Li to MoS/Al_{13}-Mont. Thus Li can not be regarded as an effective promoter for MoS/Al_{13}-Mont in the synthesis of mixed alcohols. On the other hand, the CO conversion, the selectivity for total alcohols, and the ratio of higher alcohols to methanol in product increased by adding Cs or K in the MoS/Al_{13}-Mont catalyst.

Table 9. Effect of alkali metals on MoS/Al_{13}-Mont for the synthesis of mixed alcohols at 543 K[a]

Catalyst [b]	CO Conv. (C%)	Selectivity (C%)			$C_{2+}OH$/ MeOH [c]
		Alcohols	Hydrocarbons	CO_2	
MoS/Al_{13}-Mont	34.6	33.5	43.6	22.9	1.16
Li-MoS/Al_{13}-Mont	28.8	32.8	38.9	28.3	1.21
Cs-MoS/Al_{13}-Mont	37.2	36.8	39.5	23.7	1.29
K-MoS/Al_{13}-Mont	40.5	41.9	35.8	22.3	1.40

[a] Reaction condition: $n(H_2)/n(CO)=2$; P=6 MPa; GHSV=2400h^{-1}.
[b] Mo: 15wt%, Li, Cs or K: 3wt%.
[c] Mass ratio of higher alcohols to methanol in product.

Table 10. Effect of F-T elements on K-MoS/Al_{13}-Mont for the synthesis of mixed alcohols at 543 K[a]

Catalyst [b]	CO Conv. (C%)	Selectivity (C%)			$C_{2+}OH$/ MeOH [c]
		Alcohols	Hydrocarbons	CO_2	
K-MoS/Al_{13}-Mont	40.5	41.9	35.8	22.3	1.40
Fe-K-MoS/Al_{13}-Mont	43.9	40.1	34.4	25.5	1.52
Co-K-MoS/Al_{13}-Mont	49.4	46.7	3.6	21.7	1.68
Ni-K-MoS/Al_{13}-Mont	51.8	50.5	29.3	20.2	1.78

[a] Reaction condition: $n(H_2)/n(CO)=2$; P=6 MPa; GHSV=2400h^{-1}.
[b] Mo: 15wt%, Li, Cs or K: 3wt%, Fe, Co or Ni: 5wt%.
[c] Mass ratio of higher alcohols to methanol in product.

Thus Cs and K are effective promoters for MoS/Al_{13}-Mont in the synthesis of mixed alcohols. Because K-MoS/Al_{13}-Mont showed a higher catalytic performance than that over Cs-MoS/Al_{13}-Mont, K is the best alkali metal promoter among Li, Cs, and K for the MoS/Al_{13}-Mont catalyst.

Table 10 showed the effect of F-T elements on K-MoS/Al_{13}-Mont for the synthesis of mixed alcohols at 543 K. F-T elements (Fe, Co, and Ni) have been used as promoters on MoS-based catalysts for the synthesis of mixed alcohols [35]. As shown in table 10, the CO conversion increased by adding various F-T elements in K-MoS/Al_{13}-Mont. This indicates that an interaction between F-T elements and MoS exists in the catalysts and improves the catalytic performance of K-MoS/Al_{13}-Mont. Moreover, the ratio of higher alcohols to methanol in product increased by adding F-T elements in K-MoS/Al_{13}-Mont, which implies that F-T elements mainly works in the catalyst independently for converting methanol to $C_{2+}OH$ higher alcohols. The effect of various F-T elements on K-MoS/Al_{13}-Mont are in an order of Ni > Co > Fe from both the CO conversion and the ratio of higher alcohols to methanol. Ni is the best F-T element for the K-MoS/Al_{13}-Mont catalyst in the synthesis of mixed alcohols.

Table 11 shows the catalytic performance of Ni-K-MoS/Al_{13}-Mont for the synthesis of mixed catalysts at 503 and 543 K. The CO conversion was low (14.8%) when the reaction was carried out at a low temperature of 503 K. On the other hand, the selectivity for total alcohols was high (56.4%) at 503 K because of a relatively low amount of by-product CO_2. However, because the ratio of $C_{2+}OH$ higher alcohols to methanol obtained at 503 K (0.97) was much lower than that obtained at 543 K (1.78), a high reaction temperature of 543 K was favorable for producing $C_{2+}OH$ higher alcohols over the Ni-K-MoS/Al_{13}-Mont catalyst.

Figure 7 shows the mass fractions of each alcohol in the total alcohol products over Ni-K-MoS/Al_{13}-Mont at 503 and 543 K. When the reaction was carried out at a low temperature of 503 K, the distribution of various alcohol products was closed to the A–S–F distribution because the mass fraction of each alcohol product was in an order of $CH_3OH > C_2H_5OH > C_3H_7OH > C_4H_9OH > C_5H_{11}OH > C_6H_{13}OH$. On the other hand, when the reaction was carried out at a high temperature of 543 K, the fraction of methanol decreased and ethanol became the main alcohol product over Ni-K-MoS/Al_{13}-Mont. However, the distribution of C_{2+} higher alcohols was also closed to the A–S–F distribution at a high reaction temperature of 543 K. This indicates that the reaction route over the MoS-based catalysts is similar to that over the Cu-based catalysts but is quite different from that over the Rh-based catalysts in the synthesis of alcohols from syngas.

Table 11. Catalytic performance of Ni-K-MoS/Al_{13}-Mont for the synthesis of mixed catalysts at 503 and 543 K[a]

Temperature (K)	CO Conv. (C%)	Selectivity (C%)			$C_{2+}OH$/ MeOH [b]
		Alcohols	Hydrocarbons	CO_2	
503	14.8	56.4	31.0	12.6	0.97
543	51.8	50.5	29.3	20.2	1.78

[a] Reaction conditions: $n(H_2)/n(CO)=2$; $P=3$ MPa; GHSV=2400h^{-1}; Mo: 15wt%; K: 3wt%; Ni: 5wt%.
[b] Mass ratio of higher alcohols to methanol in product.

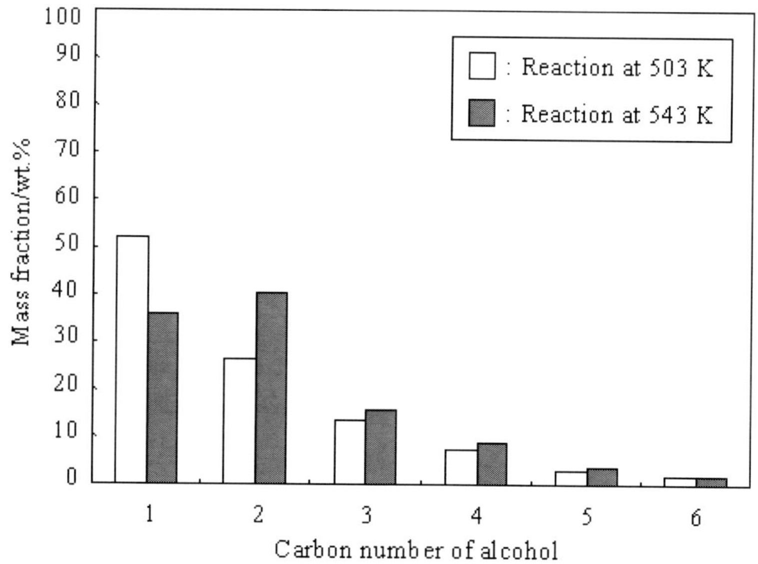

Figure 7. The mass fractions of each alcohol in the total alcohol products over Ni-K-MoS/Al_{13}-Mont at 503 and 543 K. Mo: 15 wt%, K: 3 wt%, Ni: 5 wt%. Reaction conditions: $n(H_2)/n(CO) = 2$; $P = 6$ MPa; GHSV = 2400 h^{-1}.

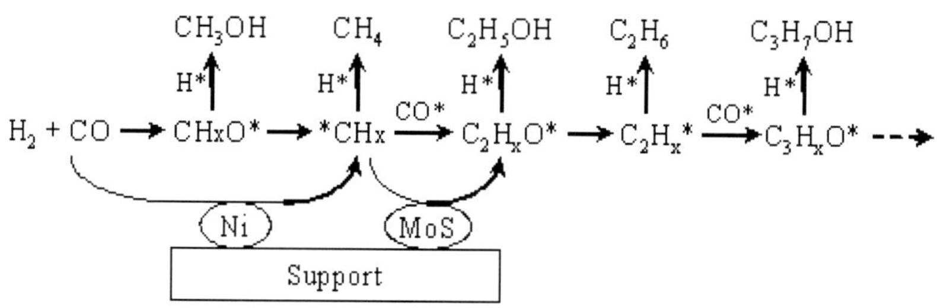

Scheme 4. Reaction pathway of CO hydrogenation over Ni-MoS-based catalysts.

Scheme 4 outlines the reaction pathway of CO hydrogenation over Ni-MoS-based catalysts. At first, CO dissociated readily to form surface CH_x* species on separated Ni and the formation of methane took place by the direct hydrogenation of CH_x* species. Then, the non-dissociated CO* which absorbed on the MoS phase inserts into a metal-methyl carbon to produce alcohol precursor.

The formed alcohol precursor further hydrogenates to form alcohols or dehydrates to form hydrocarbons [20, 38]. This route comprises both a chain growth via hydrogenation intermediates and a chain growth via oxygenated intermediates [66]. Because the chain growth can not stop after ethanol has been formed in the reaction system, The Ni-MoS-based catalysts produce mixed alcohols from syngas like the Cu-based catalysts. The fraction of ethanol in total alcohols over the Ni-MoS-based catalysts is higher than that over the Cu-based catalysts but is lower than that over the Rh-based catalysts.

Figure 8 shows the influence of reaction pressure on CO conversion and alcohols selectivity in the synthesis of alcohols from syngas over Ni-K-MoS/Al_{13}-Mont at 543 K. Because the residence time of the feed reaction gas pass through the catalyst layer increased with increasing reaction pressure, the CO conversion increased with increasing reaction pressure from 3 to 8 MPa over Ni-K-MoS/Al_{13}-Mont. A relatively sharp increase of CO conversion was observed by increasing reaction pressure from 4 to 5 MPa. On the other hand, the selectivity for total alcohols was very low at 3 MPa (12.3%) and it increased with increasing pressure in the range from 3 to 8 MPa.

Because the selectivity for total alcohols showed the highest value at 8 MPa (59.3%), a high reaction pressure is favorable for the synthesis of alcohols from syngas over Ni-K-MoS/Al_{13}-Mont.

Figure 9 shows the time sequences in the synthesis of alcohols from syngas without H_2S or containing 10 ppm H_2S over the Ni-K-MoS/Al_{13}-Mont catalyst. Using a clear feed gas without any sulfur compounds, Ni-K-MoS/Al_{13}-Mont showed an initial CO conversion of 51.8% and an initial alcohols selectivity of 50.5% at 543 K. However, the CO conversion decreased to 42.2% and the alcohols selectivity decreased to 40.9% after reaction for 72 h in the clear feed gas. Ni-K-MoS/Al_{13}-Mont was sulfided by 10% H_2S (in 90% H_2) at 400 °C prior to the reaction and a Ni-Mo-S phase was formed as the active phase for the synthesis of alcohols. The sulfur component in Ni-K-MoS/Al_{13}-Mont was lost slowly during the reaction [41]. Because the clean feed gas did not contain any sulfur components, the loss of sulfur could not be supplemented by the feedstock. As shown in figure 9, the speed of deactivation was slow in the first 10 hours and became fast after reaction for 10 hours over Ni-K-MoS/Al_{13}-Mont. The loss of sulfur caused the collapse of Ni-Mo-S active phase in Ni-K-MoS/Al_{13}-Mont. On the other hand, using a feed gas containing 10 ppm H_2S for the reaction, the CO conversion and the alcohols selectivity did not decrease after reaction for 72 hours over Ni-K-MoS/Al_{13}-Mont at 543 K. The sulfur compound (H_2S) in the feed gas supplemented the loss of sulfur from the surface of the Ni-K-MoS/Al_{13}-Mont catalyst. The Ni-Mo-S active phase was remained in the feed gas containing 10 ppm H_2S, which maintained the catalytic activity of Ni-K-MoS/Al_{13}-Mont during the reaction.

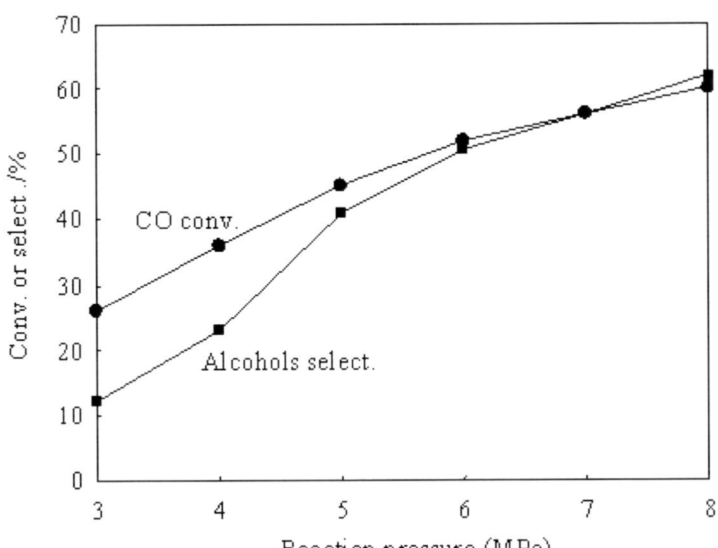

Figure 8. Influence of reaction pressure on CO conversion and alcohols selectivity in the synthesis of alcohols from syngas over Ni-K-MoS/Al_{13}-Mont at 543 K.

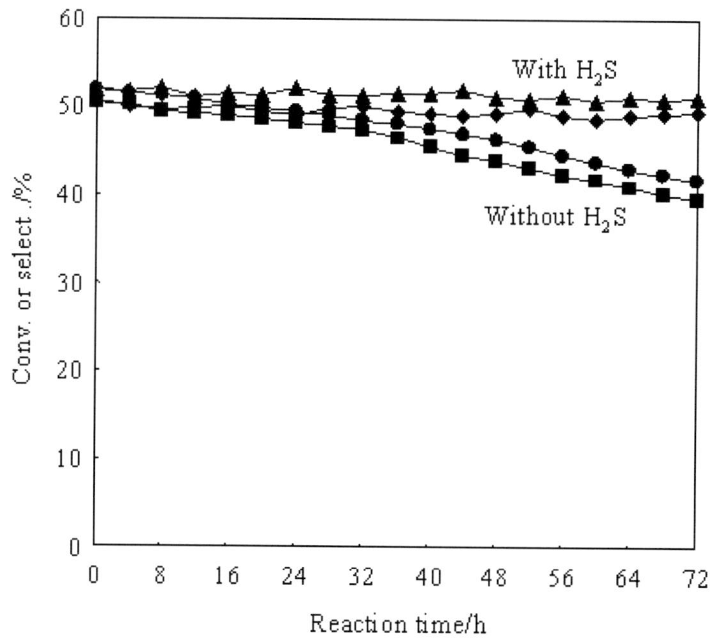

Figure 9. Time sequences in the synthesis of alcohols from syngas without H_2S or containing 10 ppm H_2S over the Ni-K-MoS/Al_{13}-Mont catalyst (●: CO conversion without H_2S; ▲: CO conversion with H_2S; ■: selectivity for alcohols without H_2S; ♦: selectivity for alcohols with H_2S).

On the whole, our Ni-K-MoS/Al_{13}-Mont catalyst showed a higher catalytic performance than those of the reported MoS-based catalysts. Ethanol was the main alcohol product but the amounts of by-products (hydrocarbons and CO_2) were relatively high over Ni-K-MoS/Al_{13}-Mont. Ni-K-MoS/Al_{13}-Mont needed a high reaction pressure for the reaction because the selectivity for alcohols greatly decreased at a reaction pressure lower than 5 MPa. We think a further reduction of hydrocarbons and CO_2 are important to achieve a large-scale commercial process using the MoS-based catalysts. The desulfurization process of the feed syngas becomes easy in using the MoS-based catalysts because they need sulfur components to maintain the catalytic stability during the reaction.

CONCLUSION

The designs of promoter and support are important to develop highly active catalysts for the synthesis of higher alcohols (including ethanol) from syngas. Alkali metals (Li, Cs, and K) and F-T elements (Fe, Co, and Ni) are effective promoters in the synthesis of higher alcohols. Both the physical property (BET surface area, porous structure, and so on) and chemical property (acidic-basic ability, reducibility, and so on) of the support influence the catalyst performance. By designing promoter and support, we developed Cs-Ni-Cu/$Ce_{0.8}Zr_{0.2}O_2$ (Cu-based catalyst), Li-Fe-Rh/$Ce_{0.8}Zr_{0.2}O_2$ (Rh-based catalyst), and Ni-K-MoS/Al_{13}-Mont (MoS-based catalyst) for the synthesis of higher alcohols from syngas in this study. The three catalysts showed higher catalytic performances than those of reported catalysts. Cs-Ni-Cu/$Ce_{0.8}Zr_{0.2}O_2$ showed the highest selectivity for total alcohols among the three catalysts but it mainly formed methanol and isobutanol from syngas. Li-Fe-

$Rh/Ce_{0.8}Zr_{0.2}O_2$ selectively formed ethanol and obtained the highest selectivity for ethanol among the three catalysts. The ethanol selectivity over Li-Fe-$Rh/Ce_{0.8}Zr_{0.2}O_2$ did not greatly decrease at low reaction pressures (1–2 MPa). Li-Fe-$Rh/Ce_{0.8}Zr_{0.2}O_2$ had a good catalytic stability in clean feed gas but deactivated when the feed gas contained 10 ppm H_2S due to the formation of inactive RhS phase. Ni-K-MoS/Al_{13}-Mont formed mixed alcohols like Cs-Ni-Cu/$Ce_{0.8}Zr_{0.2}O_2$, but the fraction of ethanol in the total alcohols over Ni-K-MoS/Al_{13}-Mont was much higher than that over Cs-Ni-Cu/$Ce_{0.8}Zr_{0.2}O_2$. A high reaction pressure (> 5 MPa) was necessary to maintain the selectivity for alcohols over Ni-K-MoS/Al_{13}-Mont. Ni-K-MoS/Al_{13}-Mont deactivated in a clean feed gas due to the loss of sulfur in the catalyst but had a good catalytic stability when the feed gas contained 10 ppm H_2S.

REFERENCES

[1] Song, C. (2006). Global challenges and strategies for control, conversion and utilization of CO_2 for sustainable development involving energy, catalysis, adsorption and chemical processing. *Catalysis Today*, *115*, 2-32.

[2] Petrus, L. and Noordermeer, M. A. (2006). Biomass to biofuels, a chemical perspective. *Green Chemistry*, *8*, 861-867.

[3] Cleveland, C., Hall, C. A. S. and Herendeen, R. A. (2006). Energy Returns on Ethanol Production, *Science*, *312*, 1746-1747.

[4] Palsson, B. O., Faith-Afshar, S., Rudd, D. F. and Lightfoot, E. N. (1981). Biomass as a Source of Chemical Feedstocks: An Economic Evaluation. *Science*, *213*, 513-517.

[5] Deluga, D. A., Salge, J. R., Schmidt, L. D. and Verykios, X. E. (2004). Renewable Hydrogen from Ethanol by Autothermal Reforming. *Science*, *303*, 993-997.

[6] Farrell, A. E., Plevin, R. J., Turner, B. T., Jones, A. D., O'Hare, M. and Kammen, D. M. (2006). Ethanol Can Contribute to Energy and Environmental Goals. *Science*, *311*, 506-508.

[7] Spivey, J. J. and Egbebi, A. (2007). Heterogeneous catalytic synthesis of ethanol from biomass-derived syngas. *Chemical Social Review*, *36*, 1514-1528.

[8] Subramani, V. and Gangwal, S. K. (2008). A review of recent literature to search for an efficient catalytic process for the conversion of syngas to ethanol. *Energy and Fuel*, *22*, 814-839.

[9] Pijolat, M. and Perrichon, V. (1985). Synthesis of alcohols from CO and H_2 on a Fe/$A1_2O_3$ catalyst at 8-30 bars pressure. *Applied Catalysis*, *13*, 321-333.

[10] Okabe, K., Yamada, H., Hanaoka, T., Matsuzaki, T., Arakawa, H. and Abe, Y. (2001). CO_2 hydrogenation to alcohols over highly dispersed Co/SiO_2 catalysts derived from acetate. *Chemistry Letters*, *30*, 904-905.

[11] Nunan, J. G., Bogdan, C. E., Klier, K., Smith, K. J., Young, C. W. and Herman, R. G. (1988). Methanol and C_2 oxygenate synthesis over cesium doped Cu/ZnO and Cu/ZnO/Al_2O_3 catalysts: a study of selectivity and ^{13}C incorporation patterns. *Journal of Catalysis*, *113*, 410-433.

[12] Nunan, J. G., Bogdan, C. E., Klier, K., Smith, K. J., Young, C. W. and Herman, R. G. (1989). Higher alcohol and oxygenated synthesis over cesium-doped Cu/ZnO catalysts. *Journal of Catalysis*, *116*, 195-221.

[13] Xu, M. T. and Iglesia, E. (1999). Carbon-carbon bond formation pathways in CO hydrogenation to higher alcohols. *Journal of Catalysis, 188*, 125-131.
[14] Hilmen, A. M, Xu, M., Gines, M. J. L. and Iglesia, E. (1998). Synthesis of higher alcohols on copper catalysts supported on alkali-promoted basic oxides. *Applied Catalysis A: General, 169*, 355-372.
[15] Sun, X. and Robert, G. W. (2003). Synthesis of higher alcohols in a slurry reactor with cesium-promoted zinc chromite catalyst in decahydronaphthalene. *Applied Catalysis A: General, 247*, 133-142.
[16] Subramanian, N. D., Balaji, G., Kumar, C. S. S. R. and Spivey, J. J. (2009). Development of cobalt-copper nanoparticles as catalysts for higher alcohol synthesis from syngas. *Catalysis Today, 147*, 100-106.
[17] Verkerk, K. A. V., Jaeger, B., Finkeldei, C. H. and Keim, W. (1999). Recent development in isobutanol synthesis from synthesis gas. *Applied Catalysis A: General, 186*, 407-431.
[18] Herman, R. G. (2000). Advances in catalysis synthesis and utilization of higher alcohols. *Catalysis Today, 55*, 233-245.
[19] Liu, Y., Murata, K., Inaba, M., Takahara, I. and Okabe, K. (2010). Synthesis of mixed alcohols from syngas over Cs- modified $Cu/Ce_{1-x}Zr_xO_2$ catalysts. *Journal of the Japan Petroleum Institute, 53*, 153-159.
[20] Fang, K., Li, D., Lin, M., Xiang, M., Wei, W. and Sun, Y. (2009). A short review of heterogeneous catalytic process for mixed alcohols synthesis via syngas. *Catalysis Today, 147*, 133-138.
[21] Du, Y., Chen, D. and Tsai, K. (1987) Promoter action of rare earth oxides in rhodium/silica catalysts for the conversion of syngas to ethanol. *Applied Catalysis, 35*, 77-92.
[22] Hu, J., Wang, Y., Cao, C., Elliott, D. C., Stevens, D. J., and White, J. F. (2007). Conversion of biomass-derived syngas to alcohols and C_2 oxygenates using supported Rh catalysts in a microchannel reactor. *Catalysis Today, 120*, 90-95.
[23] Haider, M. A., Gogate, M. R. and Davis, R. J. (2009). Fe-promotion of supported Rh catalysts for direct conversion of syngas to ethanol. *Journal of Catalysis, 261*, 9-16.
[24] Chuang, S. S. C., Stevens, Jr., R. W. and Khatri, R. (2005). Mechanism of C_{2+} oxygenate synthesis on Rh catalysts. *Topics in Catalysis, 32*, 225-232.
[25] Egbebi, A., Schwartz, V., Overbury, S. H. and Spivey, J. J. (2010). Effect of Li pronoter on titania-supported Rh catalyst for ethanol formation from CO hydrogenation. *Catalysis Today, 149*, 91-97.
[26] Burch, R. and Petch, M. I. (1992). Investigation of the synthesis of oxygenates from carbon monoxide/hydrogen mixture on supported rhodium catalysts. *Applied Catalysis A: General, 88*, 39-60.
[27] Orita, H., Naito, S. and Tamaru, K. (1983). Effect of SMSI (strong-support interaction) behaviour on $CO\text{-}H_2$ reaction over Rh catalysts. *Journal of Chemical • Society, Chemical Communications*, 993-994.
[28] Wang, Y., Luo, H., Liang, D. and Bao, X. (2000). Different mechanisms for the formation of acetaldehyde and ethanol on the Rh-based catalysts, *Journal of Catalysis, 196*, 46-55.

[29] Ehwald, H., Ewald, H., Gutschick, D., Hermann, M., Miessner, H., Ohlmann, G.. and Schierhorn, E. (1991). A bicomponent catalyst for the selective formation of ethanol from synthesis gas. *Applied Catalysis*, *76*, 153-169.

[30] Burch, R. and Hayes, M. J. (1997). The preparation and characterization of Fe-promoted Al_2O_3-supported Rh catalysts for the selective production of ethanol from syngas. *Journal of Catalysis*, *165*, 249-261.

[31] Liu, Y., Murata, K., Inaba, M., Takahara, I. and Okabe, K. (2011). Synthesis of ethanol from syngas over $Rh/Ce_{1-x}Zr_xO_2$ catalysts. *Catalysis Today*, *164*, 308-314.

[32] Ngo, H., Liu, Y. and Murata, K. (2011). Effect of secondary additives (Li, Mn) in Fe-promoted Rh/TiO_2 catalysts for the synthesis of ethanol from syngas. *Reaction Kinetics, Mechanisms and Catalysis*, *102*, 425-435.

[33] Bao, J., Sun, Z. H., Fu, Y. L., Bian, G. Z., Zhang, Y. and Tsubaki, N. (2009). Mixed alcohol synthesis from syngas on K–Co–Mo/C catalyst prepared by a sol–gel method. *Topics in Catalysis*, *52*, 789-794.

[34] Xiang, M., Li, D., Qi, H., Li, W., Zhong, B. and Sun Y. (2007). Mixed alcohols synthesis from carbon monoxide hydrogen over potassium promoted β-Mo_2C catalysts. *Fuel*, *86*, 1298-1303.

[35] Xiang, M., Li, D., Xiao, H., Zhang, J., Qi, H., Li, W., Zhong, B. and Sun, Y. (2008). Synthesis of higher alcohols from syngas over Fischer–Tropsch elements modified K/β-Mo_2C catalysts. *Fuel*, *87*, 599-603.

[36] Xiang, M., Li, D., Xiao, Li, W., Zhong, B. and Sun, Y. (2006). Performances of mixed alcohols synthesis over potassium promoted molybdenum carbides. *Fuel*, *85*, 2662-2665.

[37] Bian, G., Z., Fu, Y., L. and Ma, Y. S. (1999). Structure of Co–K–Mo/γ-Al_2O_3 catalysts and their catalytic activity for mixed alcohols synthesis. *Catalysis Today*, *51*, 187-193.

[38] Li, D., Yang, C., Li, W., Sun, Y. and Zhong, B. (2005). Ni/ADM: a high activity and selectivity to $C_{2+}OH$ catalyst for catalytic conversion of synthesis gas to C_1-C_5 mixed alcohols. *Topics in Catalysis*, *32*, 233-239.

[39] Iranmahbood, J. and Hill, D. O. (2002). Alcohols synthesis from syngas over $K_2CO_3/CoS/MoS_2$ on activated carbon. *Catalysis Letters*, *78*, 49-55.

[40] Zhang, Y., Sun, Y. and Zhong, B. (2001). Synthesis of higher alcohols from syngas over ultrafine Mo-Co-K catalysts. *Catalysis Letters*, *76*, 249-253.

[41] Li, X., Feng, L., Liu, Z., Zhong, B., Dadyburjor, D. B. and Kungler, E. L. (1998). Higher alcohols from synthesis gas using carbon-supported doped molybdenum-based catalysts. *Industrial and Engineering Chemistry Research*, *37*, 3853-3863.

[42] Iranmahboob, J., Toghiani, H. and Hill, D.O. (2003). Dispersion of alkali on the surface of Co-MoS_2/clay catalyst: A comparison of K and Cs as a promoter for synthesis alcohol. *Applied Catalysis A: General*, *247*, 207-218.

[43] Iranmahboob, J., Hill, D.O. and J., Toghiani, H. (2002). K_2CO_3/Co-MoS_2/clay catalysts for stnthesis of alcohols: influences of potassium and cobalt. *Applied Catalysis A: General*, *231*, 99-108.

[44] Liu, Y., Murata, K., Okabe, K., Inaba, M., Takahara, I., Hanaoka, T. and Sakanishi, K. (2009). Selective hydrocracking of Fischer-Tropsch waxes to high-quality diesel fuel over Pt-promoted polyoxocation-pillared montmorillonites. *Topics in Catalysis*, *52*, 597-608.

[45] Liu, Y., Sotelo-Boy, R., Murata, K., Minowa, T. and Sakanishi, K. (2012). Production of bio-hydrogenated diesel by hydrotreatment of high-acid-value waste cooking oil over ruthenium catalyst supported on Al-polyoxocation-pillared montmorillonite, *Catalysts*, *2*, 171-190.
[46] Liu, Y., Hanaoka, T., Murata, K., Okabe, K., Inaba, M., Takahara, I. and Sakanishi, K. (2007). Hydrocracking of Fischer-Tropsch wax to diesel-range hydrocarbons over bifunctional catalysts containing Pt and polyoxocation-pillared montmorillonite. *Chemistry Letters*, *36*, 1470-1471.
[47] Liu, Y., Sotelo-Boyás, R., Murata, K., Minowa, T. and Sakanishi, K. (2009). Hydrotreatment of Jatropha oil to produce green diesel over trifunctional Ni-Mo/SiO_2-Al_2O_3 catalyst. *Chemistry Letters*, *38*, 552-553.
[48] Liu, Y., Sotelo-Boyás, R., Murata, K., Minowa, T. and Sakanishi, K. (2011). Hydrotreatment of vegetable oils to produce bio-hydrogenated diesel and liquefied petroleum gas fuel over catalysts containing sulfided Ni-Mo and solid Acids. *Energy and Fuels*, *25*, 4675-4685.
[49] Bunch, A. Y., Wang, X. and Ozkan, U. S. (2007). Hydrodeoxygenation of benzofuran over sulfided and reduced Ni-Mo/γ-Al_2O_3 catalyst: Effect of H_2S. *Journal of Molecular Catalysis A: Chemical*, *270*, 264-272.
[50] Fujitani, T., Saito, M., Kanai, Y., Kakumoto, T. and Watanabe, T. (1994). The role of metal oxides in promoting a copper catalyst for methanol synthesis. *Catalysis Letters*, *25*, 271-276.
[51] Liu, Y., Hayakawa, T., Suzuki, K., Hamakawa, S., Tsunoda, T., Ishii, T. and Kumagai M. (2002). High active copper/ceria catalysts for the steam reforming of methanol. *Applied Catalysis A: General*, *223*, 137-145.
[52] Liu, Y., Hayakawa, T., Tsunoda, T., Suzuki, K., Hamakawa, S., Murata, K., Shiozaki, R., Ishii, T. and Kumagai, M. (2003). Steam reforming of methanol over Cu/CeO_2 catalysts studied in comparison with Cu/ZnO and Cu/Zn(Al)O catalysts. *Topics in Catalysis*, *22*, 205-213.
[53] Liu, Y., Hayakawa, T., Ishii, T., Kumagai, M., Yasuda, H., Suzuki, K., Hamakawa, S. and Murata, K. (2001). Methanol decomposition to synthesis gas at low temperature over palladium support on ceria-zirconia solid solutions. *Applied Catalysis A: General*, *210*, 301-314.
[54] Liu, Y., Suzuki, K., Hamakawa, S., Hayakawa, T., Murata, K., Ishii, T. and Kumagai, M. (2000). Highly active methanol decomposition catalyst derived from Pd-hydrotalcite dispersed on mesoporous silica. *Catalysis Letters*, *66*, 205-213.
[55] Zhu, Z. H. and He, D. H. (2008). CO hydrogenation to iso-C_4 hydrocarbons over CeO_2-TiO_2 catalysts. *Fuel*, *87*, 2229-2235.
[56] Khandee, W., Jongsomjit, B., Assbumrungrat, S., Praserthdam, P. and Goto, S. (2007). Investigation of isosynthesis via CO hydrogenation over ZrO_2 and CeO_2 catalysts: Effect of crystallite size, phase composition and acid-base sites. *Catalysis Communications*, *8*, 548-556.
[57] Tsubaki, N. and Fujimoto, K. (2003). Promotional SMSI effect on supported palladium catalysts for methanol synthesis. *Topics in Catalysis*, *22*, 325-335.
[58] Shen, W. J., Ichihashi, Y. and Matsumura, Y. (2002). A comparative study of palladium and copper catalysts in methanol synthesis. *Catalysis Letters*, *79*, 125-127.

[59] Takeuchi, A. and Katzer, J. R. (1982). Ethanol formation mechanism from carbon monoxide + molecular hydrogen. *The Journal of Physical Chemistry*, *86*, 2438-2441.

[60] Gurbus, E. I., Kunkes, E. L. and Dumesic, J. A. (2010). Integration of C–C coupling reactions of biomass-derived oxygenates to fuel-grade compounds. *Applied Catalysis B: Environmental*, *94*, 134-141.

[61] Gurbus, E. I., Kunkes, E. L. and Dumesic, J. A. (2009). Vapour-phase C–C coupling reactions of biomass-derived oxygenates over Pd/$CeZrO_x$ catalysts. *Journal of Catalysis*, *266*, 236-249.

[62] Mott, C. J. B. (1988). Clay minerals—An introduction. *Catalysis Today*, *2*, 199-208.

[63] Kloprogge, J. T. (1998) Synthesis of smectites and porous pillared clay catalysts: A review. *Journal Porous Materials*, *5*, 5-41.

[64] Martinez-Oritz, M. J., Fetter, G., Dominguez, J. M., Melo-Banda, J.A. and Ramos-Gomez, R. (2003). Catalytic hydrotreating of heavy vacuum gas oil on Al- and Ti-pillared clays prepared by conventional and microwave irradiation methods. *Microporous and Mesoporious Materials*, *58*, 73-80.

[65] Kloprogge, J. T., Welters, W. J. J., Booy, E., Beer, V. H. J., Santen, R. A., Geus, J. W. and Jansen, J. B. H. (1993). Catalytic activity of nickel sulfide catalysts supported on Al-pillared montmorillonite for thiophene hydrodesulfurization, *Applied Catalysis A: General*, *97*, 77-85.

[66] Chuang, S. C., Tian, Y. H., Good win, Jr. J. G. and Wender, I. (1985). The use of probe molecules in the study of CO hydrogenation over SiO_2-supported Ni, Ru, Rh, and Pd. *Journal of Catalysis*, *96*, 396-407.

In: Ethanol: Production, Cellular Mechanisms and Health Impact ISBN: 978-1-62257-297-7
Editors: Roy I. Henry and Billy P. Woods

Chapter 3

INORGANIC MEMBRANE REACTOR FOR HYDROGEN PRODUCTION THROUGH BIO-ETHANOL REFORMING PROCESSES

A. Iulianelli[1], S. Liguori[1], P. Pinacci[2], P. Morrone[3] and A. Basile[*,1]

[1]ITM-CNR, University of Calabria, Rende (CS), Italy
[2]RSE SpA, Milano (MI), Italy
[3]Mechanical Engineering Dpt. of University of Calabria, Rende (CS), Italy

ABSTRACT

Today, hydrogen is considered as the most viable energy carrier useful for many applications and, particularly, for fuel cells supplying. Taking into account the environmental contamination caused by the exploitation of derived of fossil fuels, hydrogen production from ethanol coming from biomass as a renewable source could constitute an interesting and strategic opportunity. This chapter describes the possibility of producing hydrogen from ethanol reforming processes through inorganic membrane reactors. The latter are proposed and described because useful to perform reactions limited by the equilibrium conversion because of the hydrogen selective separation capability of the inorganic membranes. Both benefits and drawbacks of the inorganic membrane reactor technology are examined and, when applied to reforming reactions, their performances such as hydrogen yield, hydrogen recovery, ethanol conversion and so on are emphasized and qualitatively compared to those of the conventional systems. Furthermore, a short overview from the open specialized literature on the most representative scientific results concerning ethanol reforming processes performed by inorganic membrane reactor technology is also given and discussed.

Keywords: Hydrogen production, membrane reactors, ethanol steam reforming, inorganic membranes

---

[*] Corresponding author: a.basile@itm.cnr.it, Phone: +39 0984 492013; Fax: +39 0984 402103.

INTRODUCTION

During the last thirty years or more, several actions culminated to Kyoto Protocol were oriented to the depletion of atmospheric pollution and greenhouse gas emissions, with a consequent development of new technologies for mitigating these phenomena. Among other technologies, proton exchange membrane fuel cells (PEMFCs) represent one of the most interesting candidates to cover this need. As well known, PEMFCs are supplied by high pure hydrogen, which is industrially produced at moment as a hydrogen rich-stream mainly via steam reforming of natural gas [1] and further purified to achieve the desired purity via conventional systems. Therefore, the growing attention towards hydrogen production was also testified at scientific level, as reflected in the open literature by the increasing number of scientific publications per year, Figure 1.

However, as confirmed by the International Energy Agency, today the world economy is still devoted to the exploitation of derived fossil fuels such as oil, coal, methane, etc. [2]. Therefore, taking into account the increase of environmental pollution due to CO_2 emissions and other greenhouse gases caused by burning of fossil fuels, the development of new technologies and the exploitation of renewable sources represent a top priority. Thus, the aforementioned fuel cells could represent one of the most promising technologies for matching such objectives as applications to large-scale stationary systems as well as small-scale portable power supplying devices [3].

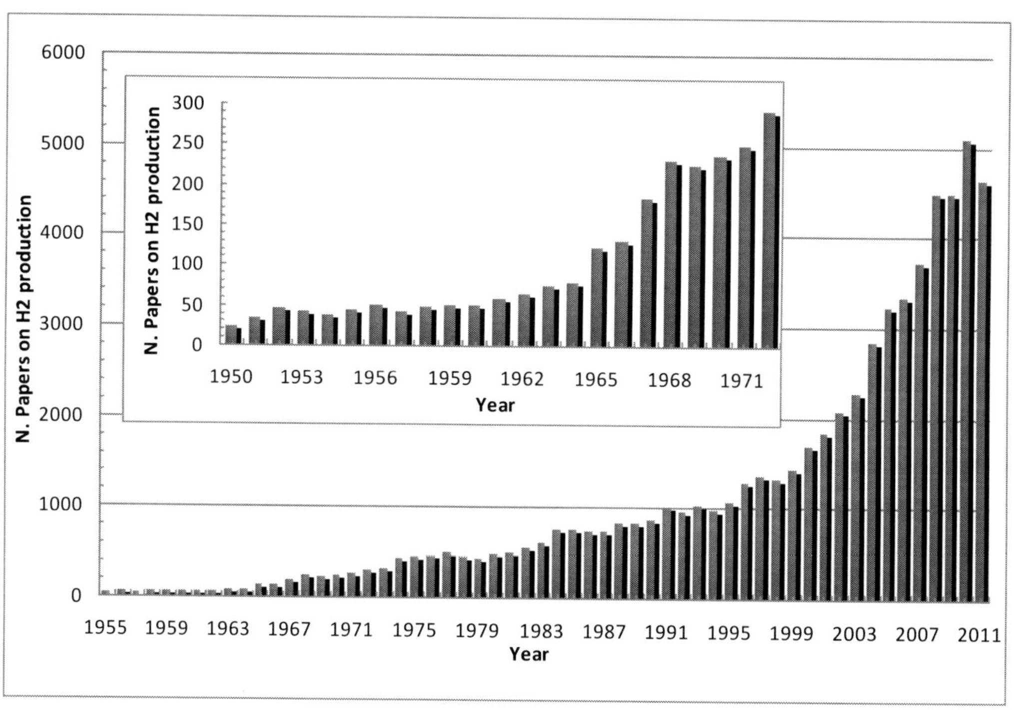

Figure 1. Number of scientific publications on H_2 production vs year. Scopus database: www.scopus.com.

In particular, PEMFCs transform chemical energy produced by electrochemical reaction between hydrogen and oxygen into electrical energy without pollutants besides thermal energy given by the direct combustion of hydrogen and oxygen. Furthermore, PEMFCs produce more power for a given volume or weight of fuel cell with respect to other kind of fuel cells, making them compact and lightweight. Generally, they are performed at temperature lower than 100 °C, allowing a rapid start-up. These characteristics make PEMFCs as top candidates for an alternative technology to the conventional systems. Nevertheless, PEMFCs need a highly pure hydrogen supplying because of the anodic Pt catalysts tolerance of only a few ppm of CO (<10).

Industrially, hydrogen is produced via steam reforming reaction of fossil fuels such as natural gas, etc. It is conventionally performed via fixed bed reactors (FBRs) and the hydrogen rich-stream coming out from the conventional reformer contains besides hydrogen also other byproduct gases such as mainly CO, CH_4 and CO_2. Therefore, for PEMFC supplying, hydrogen needs to be purified by means of further processes like water gas shift (WGS) reaction, pressure swing adsorption (PSA) and/or Pd membrane separation, etc. In this contest, membrane reactor (MR) technology could play a key role and constitute an alternative solution to the conventional systems, offering the opportunity of combining the reforming reaction for producing hydrogen and its separation/purification in the same device [4].

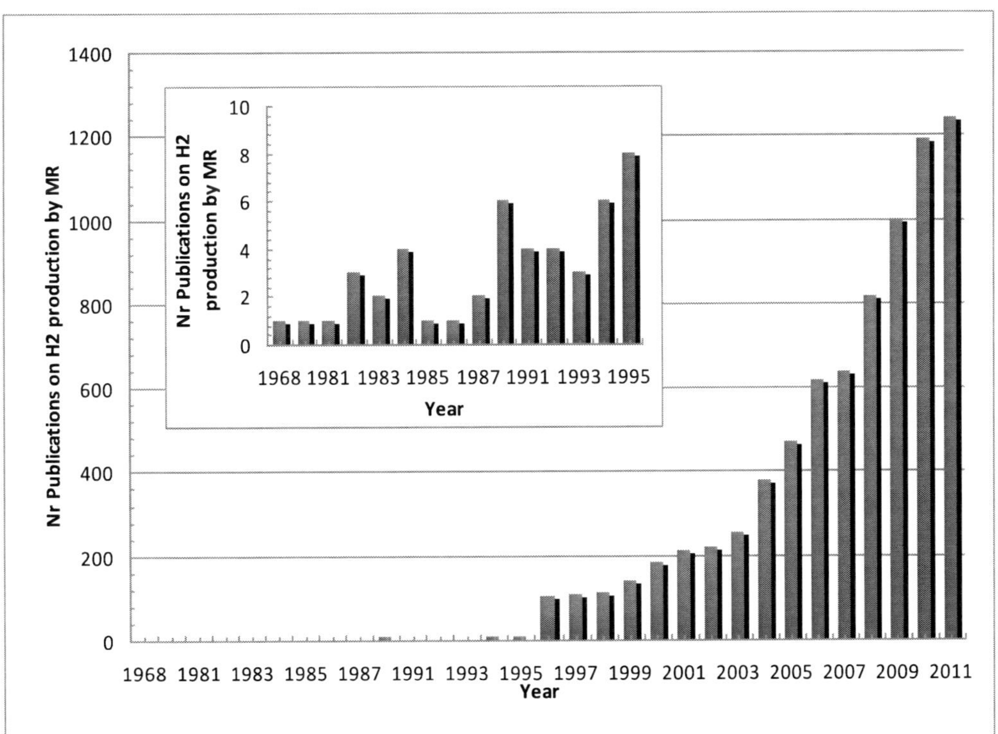

Figure 2. Number of scientific papers on H_2 production by MR technology vs year. Scopus database: www.scopus.com.

However, only a few applications on MR technology for hydrogen production at industrial scale can be noticed (e.g. Tokyo Gas Company Ltd., http://www.tokyo-gas.co.jp/index_e.html), while at scientific level many efforts in terms of scientific studies have been done in the area for proposing MR technology at larger scale, Figure 2.

In order to give some highlights on the advantages of MR technology over the FBRs, the benefits due to the inorganic MRs utilization are summarized in the following [5,6]:

- chemical reaction and hydrogen separation performing in only one system with a consequent reduction of the capital costs;
- conversion enhancement of equilibrium limited reactions;
- higher conversions than FBRs (operated at the same MRs conditions) or, on the contrary, the same conversion, but operating at milder conditions than FBRs;
- improvement of both hydrogen yield and hydrogen selectivity;
- direct production of high purity hydrogen (in the case of dense Pd-based MRs).

Among several kind of inorganic membranes, particular attention should be devoted to the dense palladium-based membranes because of their full hydrogen perm-selectivity [7]. The hydrogen permeation through dense Pd-based membranes follows a solution-diffusion transport mechanism [8]. However, when at temperatures lower than 300 °C a dense palladium membrane is exposed to a hydrogen flow, the embrittlement phenomenon takes place after a few cycles of $\alpha \leftrightarrows \beta$ transitions of pure palladium [9-11]. These transitions do not take place as a change of the lattice structure, but only in a lattice dilatation. As a solution of this problem, by alloying palladium with elements such as silver, copper or other metals, Pd-H phases show an increased reticular step and the ability of anticipating the reticular expansion from hydrogen [12].

Since 1960s, the behaviour of dense Pd-based and porous membranes has been studied when involved in MRs applications for hydrogen production. Nevertheless, the superiority of palladium membranes overcomes that of all the other materials owing to their high solubility of hydrogen and the aforementioned full perm-selectivity to hydrogen permeation with respect to all of other gases.

Then, the objective of this chapter is to point out the progress at scientific level concerning the reforming processes via inorganic MRs technology by exploiting ethanol as a renewable source with the intent of matching an important goal in the area of hydrogen production, in the meanwhile offering several advantages over the current technology based on conventional systems and derived fossil fuel sources.

1. Hydrogen Production From Bio-Sources: Ethanol Reforming in Fixed Bed Reactors

Among several liquid fuels, such as methanol, glycerol, acetic acid, diethyl ether, etc., ethanol represents an important bio-source and seems to be more suitable owing to its very low toxicity [13]. Generally, bio-ethanol indicates an aqueous solution containing between 8 and 12 wt% of ethanol and other by-products [14-16].

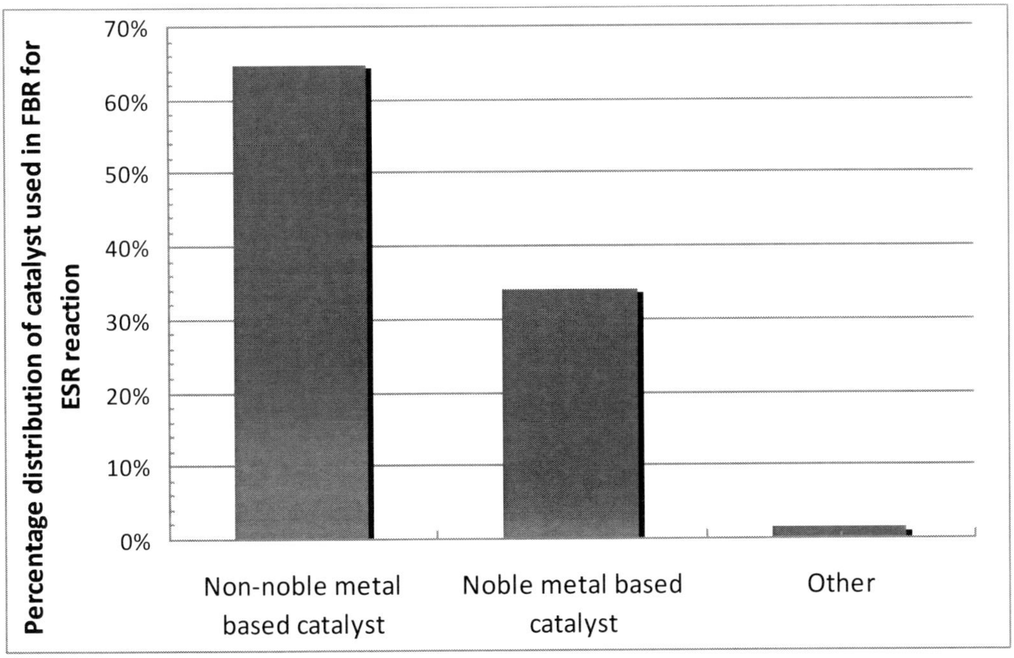

Figure 3. Percentage distribution of the catalyst types used in scientific studies focusing on ESR reaction in FBRs. Database: http://scienceserver.cilea.it.

Conventionally, a hydrogen rich-stream can be produced by converting pure ethanol mixed to steam or directly bio-ethanol via the autothermal reforming, steam reforming and oxidative steam reforming reactions. In the open literature, most of the scientific publications deals with performing ethanol steam reforming (ESR) reaction in FBRs. Figure 3 shows the percentage distribution of the catalysts mainly used for ESR reaction in FBRs. This figure highlights that non-noble metal catalysts are more commonly used, probably because the intent of many researches is not only devoted to maximize the hydrogen production but also to take an economical advantage by operating with cheaper catalysts than noble-metal ones.

In the open literature, several works focus on ESR carried out in FBRs, emphasizing the performances in terms of ethanol conversion, hydrogen yield and hydrogen selectivity by using various catalysts [17-22]. Then, the role of the catalyst is quite relevant and, in this contest, Figure 4 shows the percentage distribution of different metal phases in both non-noble metal and noble metal catalysts utilized in several scientific studies.

In particular, the current state of the art on ESR reaction in FBRs may be resumed by two interesting reviews [17,18], in which a comparative analysis on the performances using different catalysts such as Rh, Ru, Pd, Pt, Ni, Co and Cu-based is given. In these reviews, it was pointed out that FBRs performances may vary greatly, depending on the catalyst choice and reaction conditions. Nevertheless, Haga et al. [19] affirm that Co-Al_2O_3 catalyst seems to be the most selective towards ESR reaction. Furthermore, other authors support the role of cobalt as the most effective catalyst for ESR reaction because of its high catalytic activity and because it is more cost effective than noble metal-based catalysts like Rh, Pt and Pd [20-22].

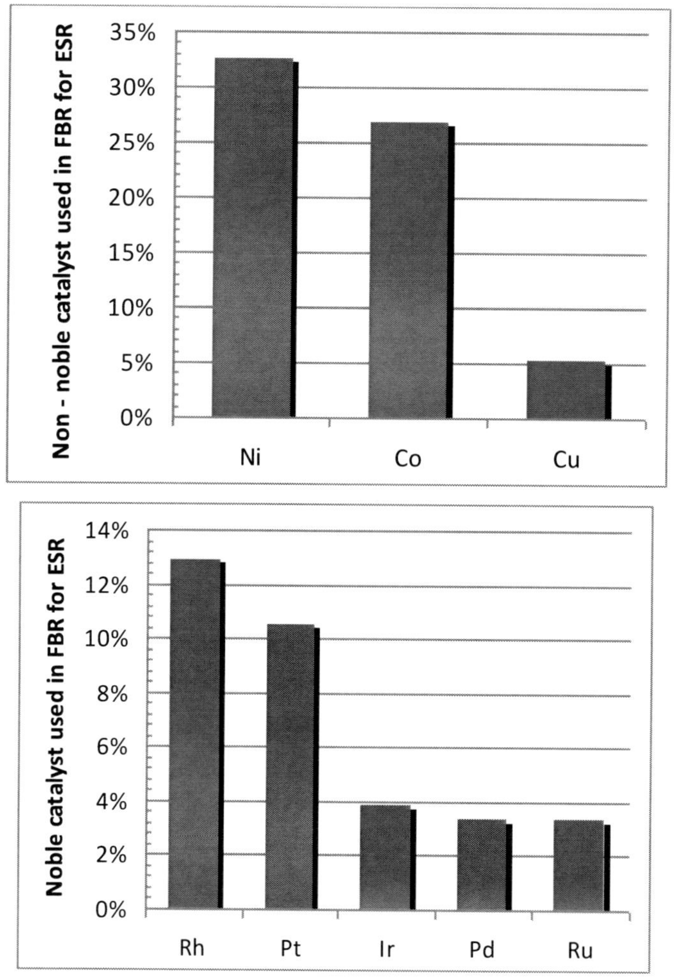

Figure 4. Percentage distribution of noble and non-noble metal catalysts in scientific studies focusing on ESR reaction performed in FBRs. Database: http://scienceserver.cilea.it.

From a thermodynamic point of view, it is well known that a water excess during ESR reaction (1) improves the conversion and reduces the formation of carbon monoxide in the reformed stream.

$$C_2H_5OH + 3H_2O = 2CO_2 + 6H_2 \qquad \Delta H^\circ_{298\,K} = 157.0\ kJ/mol \qquad (1)$$

Therefore, it is an interesting and advantageous approach to supply directly into a FBR a bio-ethanol mixture (corresponding to a water/ethanol feed molar ratio between 29.0/1 – 18.7/1), without making any distillation of further ethanol separation/purification process. This approach could be economically relevant, particularly taking into account that, in the ethanol purification process, extracting 99% water presents high cost owing to the ethanol/water azeotrope.

In this contest, Ni et al. [15] performed bio-ethanol steam reforming (BESR) reaction in FBRs, reviewing the crucial role played by the catalyst choice in terms of hydrogen production and concluding that Rh and Ni-based are the most commonly used catalysts for

BESR reaction. Moreover, other authors paid special attention to both autothermal and oxidative BESR in FBRs to highlight viable methods for improving the hydrogen yield and lowering the CO content in the reformed streams [23-25].

However, ESR reaction shows a complex reaction system producing undesirable byproducts besides hydrogen, depending on the catalyst choice [26]. Referring to Co-based catalysts, methane as a byproduct is produced by decomposition reaction (2), which takes place at temperature higher than 200 °C, or by methanation reaction (3), occurring only when Co is supported on SiO_2, MgO, ZrO [19].

$$C_2H_5OH = CH_4 + CO + H_2 \qquad \Delta H^\circ_{298\,K} = 33.16 \text{ kJ/mol} \qquad (2)$$

$$CO + 3H_2 = CH_4 + H_2O \qquad \Delta H^\circ_{298\,K} = -206.2 \text{ kJ/mol} \qquad (3)$$

Furthermore, Song et al. [14] proposed other possible reactions taking place during ESR process as reported in the following:

$$C_2H_5OH = CH_3CHO + H_2 \text{ (dehydrogenation)} \qquad \Delta H^\circ_{298\,K} = 17.36 \text{ kJ/mol} \qquad (4)$$

$$C_2H_5OH = C_2H_4 + H_2O \text{ (dehydration)} \qquad \Delta H^\circ_{298\,K} = 28.99 \text{ kJ/mol} \qquad (5)$$

$$C_2H_5OH + H_2O = 2CO + 4H_2 \text{ (incomplete reforming)} \qquad \Delta H^\circ_{298\,K} = 239.31 \text{ kJ/mol} \qquad (6)$$

$$2C_2H_5OH = (C_2H_5)_2O + H_2O \text{ (dehydrative coupling)} \qquad \Delta H^\circ_{298\,K} = -88.48 \text{ kJ/mol} \qquad (7)$$

$$CH_4 + 2H_2O = CO_2 + 4H_2 \text{ (methane steam reforming)} \qquad \Delta H^\circ_{298\,K} = 206.2 \text{ kJ/mol} \qquad (8)$$

$$CO + H_2O = CO_2 + H_2 \text{ (water gas shift)} \qquad \Delta H^\circ_{298\,K} = -41.02 \text{ kJ/mol} \qquad (9)$$

$$2CO = CO_2 + C \text{ (Boudouard reaction)} \qquad \Delta H^\circ_{298\,K} = -172.46 \text{ kJ/mol} \qquad (10)$$

Indeed, a typical ESR reformed stream can contain such byproducts as: CO_2, CO, CH_4, acetaldehyde, ethylene, ethane, etc. Having in mind to produce high purity (or, at least, a CO_x-free) hydrogen for direct PEMFC supplying, the hydrogen rich-stream coming out from a FBR needs to be separated/purified by means of further processes. Hence, considering that it would economically hopeful to limit the number of operation units for the hydrogen separation/purification, in the membranologists' area many researchers propose to replace the conventional system (reformer + gas cleaning units) with a MR, able to perform both the reaction and the hydrogen separation/purification processes in the same device [27].

However, in the following paragraph, the MR technology will be summarized, highlighting the main characteristics of the inorganic membrane reactors useful for hydrogen production.

2. Inorganic Membranes and Membrane Reactors Outlines

As a general consideration, the membranes are classified depending on their nature, geometry and separation regime [28]. Furthermore, IUPAC [29] defines membranes as a material having lateral dimensions much higher than its thickness, through which a mass transfer may take place as a consequence of a driving force occurring as a gradient of concentration, pressure, temperature, electric potential, etc.

The classification by nature subdivides the membranes into biological and synthetic. Biological membranes are manufactured easily, although they present relevant drawbacks related to the limited operating temperature (below 100 °C), pH range, etc. [30]. Synthetic membranes can be further classified into organic (polymeric) and inorganic (ceramic, metallic) depending on the operative temperature limit. Indeed, polymeric membranes generally operate up to 200 °C, preferentially under 100-150 °C, making them unusable for reforming process in MRs.

Otherwise, inorganic membranes operate at T > 250 °C and are stable between 300 - 800 °C. In some cases, they can operate at elevated temperatures (ceramic membranes) over 1000 °C [31]. However, they may be further subdivided into porous and non porous (or dense). Porous membranes are classified according to their pore diameter in microporous (d_p < 2nm), mesoporous (2nm < d_p < 50nm) and macroporous (d_p > 50nm) [28], while dense membranes are characterized by the absence of pores.

An important type of dense membrane class is represented by the metallic membranes, which can be categorized further into supported and unsupported. There are different separation mechanisms that are based on the specific properties of the materials [4] and determine the classification of the membranes, respectively depending on the different occurring mechanisms such as porous and dense. However, based on the dimension of the membrane pores, Table 1 summarizes the mechanism taking place in porous and dense membranes.

Table 1. Transport mechanisms in both porous and dense membranes

Membrane type	ϕ_{pore} (nm)	Transport mechanism
Macroporous	>50	Poiseuille (Viscous flow)
Mesoporous	2-50	Knudsen
Microporous	<2	Activated process
Dense (Pd)	-	Fick

Inorganic membranes can be also subdivided depending on the structural material as reported below:

- metal membranes,
- ceramic membranes,
- carbon membranes,
- zeolite membranes.

Metal membranes are conventionally studied for hydrogen separation from gas mixtures and in MR field [4,7,9]. Palladium and its alloys are the dominant materials for preparing this kind of membranes owing to their high solubility and permeability to hydrogen. Nevertheless, other dense membranes are selectively permeable only to hydrogen such as those based on tantalum, vanadium, nickel and titanium. Furthermore, from an economic point of view, they are less expensive than the palladium and its alloy. Unfortunately, metal membranes present as an important drawback the surface poisoning, more significant when they are very thin.

Ceramic membranes are generally manufactured starting from aluminium, titanium or silica oxides [4]. They are chemically inert and stable at high temperatures and, then, particularly useful for food, biotechnology and pharmaceutical applications as well as for gas separation and in MR applications.

Carbon membranes are identified as promising candidates for gas separation as well as for application in MRs [32]. This kind of membrane is constituted by porous solids containing constricted apertures approaching the molecular dimensions of diffusing gas molecules, which are separated through molecular sieving.

Zeolites are microporous crystalline alumina-silicate with an uniform pore size and their applications are mainly devoted for uses as catalysts or adsorbents in a form of micron or submicron-sized crystallites embedded in millimeter-sized granules [33]. When manufacturing zeolite membranes, they show as a main drawback low gas fluxes compared to other inorganic membranes.

2.1. Membrane Reactors: Extractor or Distributor Modality

At industrial level, several heterogeneous gas–solid catalytic processes, conventionally performed by means of fixed, fluidised or trickle bed reactors, are performed as a combination of operations at high temperatures and in chemically harsh ambient. For this concern, inorganic membranes seem to be more adequate than polymeric ones as useful materials for MRs. Commonly, the separation behaviour of an inorganic membrane housed in a MR is advantageously used for improving the performances of a catalytic system. Then, it is possible to distinguish two main generic MR approaches such as the selective product separation (Extractor) and the selective reactant addition (Distributor) [34]. The first MR approach makes possible the in–situ removal of one of the reaction products. As an example, when a steam reforming reaction is performed in a conventional reactor, hydrogen yield and CO_2 product selectivities are limited by thermodynamics. Using a MR in an Extractor approach, the selective removal of hydrogen from the reaction side allows to overcome the thermodynamic equilibrium restrictions. This potentiality is called "shift effect" that allows to realize both higher hydrogen yields and ethanol conversions with respect to such a conventional reactor. Furthermore, this effect allows operation at milder reaction conditions in terms of temperature and pressure than the FBRs [35].

The second MR approach involves the use of the membrane to control the contact within reactants. Both full perm-selective and non full perm-selective membranes can be useful to supply distributively one of the reactants in the reaction zone. As an example, in partial oxidation reactions performed in FBRs, on one side oxygen rich-feed results in low product selectivities and high reactant conversions. On the other side, low oxygen content means high product selectivities but lower conversions. Using a membrane for distributive oxygen supplying in the reaction zone, both high conversions and product selectivities can be reached

[36,37]. Furthermore, since the reactants and oxygen are not premixed, mixtures are avoided and the flame back firing into the feed are considerably lowered.

2.2. The Perm-Selective Characteristics of Palladium-Membranes

The aforementioned "shift effect" is the key factor of a MR when applied to a such reforming process. It allows to overcome the thermodynamic equilibrium restrictions of a FBR and, in the meanwhile, it represents the main characteristic and benefit of MRs to improve the reaction conversion, hydrogen yield, etc. Depending on the difference of the inorganic membranes housed in MRs, the specialized literature often describes MR performances in terms of conversion as a function of the parameter "H", defined as permeation rate to reaction rate ratio [38].

For example, for a FBR the value H is equal to 0 because obviously no permeation takes place. Otherwise, at $H > 0$, different types of MRs are considered. In the case of low H values (i.e. low permeation to reaction rate ratio) microporous, dense and mesoporous MRs show the same behaviours [38]. At higher H values, it is possible to appreciate the differences of the different kinds of MRs.

MRs with a finite ideal permeation selectivity (representing the ratio between the permeance - as a ratio between the permeating flux and the transmembrane pressure - of a certain gas over that of a gas assumed as a blank, generally an inert as He, Ar or N_2) achieve an optimum in terms of permeability/reaction rate. Above the optimum, the reactant loss caused by permeation induces a detrimental effect on the conversion. On the contrary, higher ideal permeation selectivity corresponds to higher conversions.

MRs showing infinite ideal permeation selectivity, for example for hydrogen through dense Pd-based membranes, do not give any drawback in terms of conversion because loss of reactants do not occur.

In the field of inorganic membranes, dense metallic membranes have attracted great interest of many researchers [39]. Therefore, special attention has been devoted to dense palladium membranes owing to their full hydrogen perm-selectivity with respect to all the other gases. However, other metals such as niobium, vanadium and tantalum show higher hydrogen permeability than palladium in the range of 0 – 700 °C, although they have a stronger surface resistance to hydrogen transport than palladium. Therefore, dense palladium membranes are more considered, although their commercialization is limited by some drawbacks such as the low hydrogen permeability and high costs.

2.2.1. Hydrogen Permeation Through Pd-Based Membranes

The hydrogen molecular transport in palladium membranes takes place through a solution/diffusion mechanism and, as indicated by Dittmeyer et al. [40], it involves six different activated steps:

1) adsorption of molecular hydrogen on the palladium surface,
2) dissociation of molecular into atomic hydrogen,
3) dissolution of atomic hydrogen into the palladium matrix,
4) diffusion of atomic hydrogen through the membrane,

5) re-combination of atomic hydrogen to form hydrogen molecules at the gas/metal interface,
6) desorption of hydrogen molecules.

Each of these steps may control hydrogen permeation through dense film as a dependence on temperature, pressure, gas mixture composition and membrane thickness [41]. As a result, the hydrogen flux permeating through the membrane can be expressed as reported below:

$$J_{H_2} = Pe_{H_2} (p^n_{H_2,retentate} - p^n_{H_2,permeate})/\delta \qquad [mol/m^2 \cdot h] \qquad (11)$$

where J_{H2} is the hydrogen flux permeating through the membrane, *Pe* the hydrogen permeability, δ the membrane thickness, $p_{H2\text{-}retentate}$ and $p_{H2\text{-}permeate}$ the hydrogen partial pressures in the retentate (reaction side) and permeate (side in which hydrogen permeating through the membrane is collected) zones, respectively, n (variable in the range 0.5 - 1) the dependence factor of the hydrogen flux to the hydrogen partial pressure. The membrane thickness is an important parameter because as thin the membrane as high the hydrogen permeability, also affecting the membrane mechanical resistance. Therefore, for real applications to ensure the mechanical resistance and strength of the membrane, thicker membranes are necessary, with a consequent hydrogen permeability decrease. Nevertheless, when the pressure is relatively low, the rate-limiting step of the hydrogen permeation through the membrane is assumed to be the diffusion [42,43]. When the membrane thickness is greater than 5 μm, the equation (11) becomes Sieverts-Fick law (12):

$$J_{H_2,Sieverts\text{-}Fick} = Pe_{H_2} \cdot (p^{0.5}_{H_2,retentate} - p^{0.5}_{H_2,permeate})/\delta \qquad [mol/m^2 \cdot h] \qquad (12)$$

Otherwise, when the hydrogen-hydrogen interactions in the palladium bulk are not negligible owing to the high pressures, n becomes equal to 1:

$$J_{H_2} = Pe_{H_2} \cdot (p_{H_2,retentate} - p_{H_2,permeate})/\delta \qquad [mol/m^2 \cdot h] \qquad (13)$$

The hydrogen permeability dependence on the temperature can be expressed by an Arrhenius equation as:

$$Pe_{H_2} = Pe^0_{H_2} \, exp\,(-E_a/RT) \qquad [mol \cdot m/Pa \cdot m^2 \cdot h] \qquad (14)$$

where Pe_0 is the pre-exponential factor, E_a the apparent activation energy, R the universal gas constant and T the absolute temperature.

Therefore, when Sieverts-Fick law lives, the hydrogen flux is expressed as indicated by the Richardson's equation (15):

$$J_{H_2} = Pe^0_{H_2} \, [exp\,(-E_a/RT)] \cdot (p^{0.5}_{H_2,retentate} - p^{0.5}_{H_2,permeate})/\delta \qquad [mol/m^2 \cdot h] \qquad (15)$$

By considering that aforementioned behaviors of Pd-based membranes, several scientific studies deal with proposing as economically more advantageous method to develop a single system combining the reaction for hydrogen production and its purification as a pure stream [41,44-46]. According to the scheme proposed in Figure 5, the conventional system for pure hydrogen production could be replaced by a dense palladium MR, useful to perform the reaction as well as the hydrogen separation/purification in only one tool, representing a cost-effective solution owing to a lower number of devices.

2.3. Palladium-Based MRs

Generally, Pd-based MRs are assembled with supported and laminated palladium membranes. Palladium-based supported membranes possess a thin dense layer of palladium or its alloy, deposited onto porous supports such as porous Vycor glass (silica gel), SiO_2, Al_2O_3, and B_2O_3. Unlike Vycor glass, the other porous support give an excellent adherence and anchor effect [52].

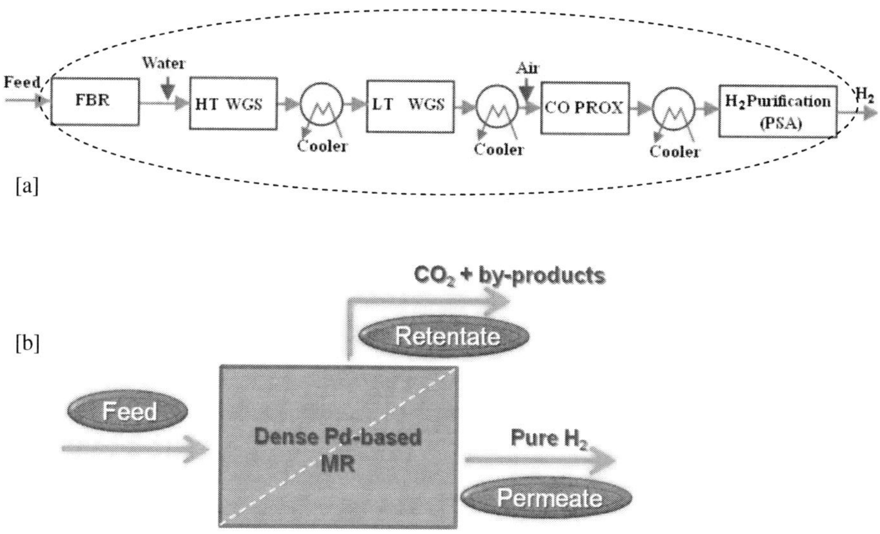

Figure 5. Conventional system (a) and dense Pd-based MR (b) for pure H_2 production from reforming reaction.

Another important support is represented by the porous stainless steel (PSS) because it shows mechanical durability, a thermal expansion coefficient close to that of palladium and the easiness of gas sealing [52]. Unfortunately, PSS support alloys the palladium at relatively high temperatures, leading to lower the hydrogen permeability [53].

Thin palladium or palladium alloys laminated membranes avoid the formation of oxides on the metallic surfaces, depleting the hydrogen adsorption activation energy and, consequently, emphasizing the hydrogen permeation behaviors. Generally, palladium alloys give some advantages over pure palladium membranes. Besides to avoid the embrittlement phenomenon, certain palladium alloys improve the membrane chemical resistance. As an example, Pd-Cu and Pd-Au alloyed membranes are more resistant to H_2S poisoning [53].

2.3.1. The Poisoning Effect on Palladium-Based Membranes

The MRs assembled with dense Pd-based membranes could show critical drawbacks related to various poisoning effects. Pure palladium membranes can be affected by "hydrogen embrittlement" phenomenon, taking place below 300 °C and 2.0 MPa over a constant hydrogen flux through the membrane. The embrittlement can be avoided by alloying palladium with other metals, commonly with silver, which displays its electron donating behaviour, being largely similar to the one of the hydrogen atom in palladium, making possible the competition for the filling of electron holes within silver and hydrogen [6].

Other negative poisoning effects for the Pd-based MRs are represented by the palladium surface contamination owing to the presence of hydrogen sulfide, SO_2, Hg vapour, thiophene, arsenic, unsaturated hydrocarbons, chlorine carbon from organic materials, etc.

- The poisoning related to the exposure to hydrogen sulphide affects negatively Pd-coated membranes, which could be destroyed rapidly and irreversibly [47,48].
- Supplying CO in a MR, a decrease of hydrogen permeation through the membrane is expected, because the adsorbed CO displaces the adsorbed hydrogen, blocking the hydrogen adsorption sites. This reduction is more significant at low temperature (below 150 °C) or at high CO feed concentration [49].
- The poisoning effect of water vapour on hydrogen permeability may be more relevant than the CO effect. It involves in the water vapour dissociation /recombinative desorption, which contaminates the palladium surface with adsorbed oxygen [50].
- When assembling a thin palladium membranes in a MR, the presence of coke on the membrane surface at high temperature could deplete the hydrogen permeation characteristics. This poisoning is caused by the carbon atoms that penetrate into the palladium lattice, provoking the membrane failure owing to the expansion of the palladium lattice [51].

3. Ethanol Reforming Reactions in MRs: A Short Overview

As previously mentioned in the first part of this chapter, during the last decades MR technology has been developed having in mind to produce hydrogen based on the exploitation of ethanol as a renewable source. Figure 6 illustrates the number of scientific studies per year focusing on hydrogen production via ESR reaction combined to MR technology with respect to the total number of publications on this field, comprising both FBR and MR applications.

Furthermore, Figure 7 points out the percentage distribution of the principal catalysts for ethanol reforming reactions in MRs. Ni and Ru-based catalysts are particularly used, probably because they are cheaper than other catalysts. Nevertheless, the intent of the scientists involved on this field is focused on emphasizing the key role of the membrane, by analyzing the performances of the reaction system in terms of ethanol conversion, hydrogen yield (defined as the ratio within hydrogen produced during the reaction and that theoretically producible from the stoichiometry of the reaction) and hydrogen recovery (hydrogen collected on the permeate side over the total hydrogen produced during the reaction).

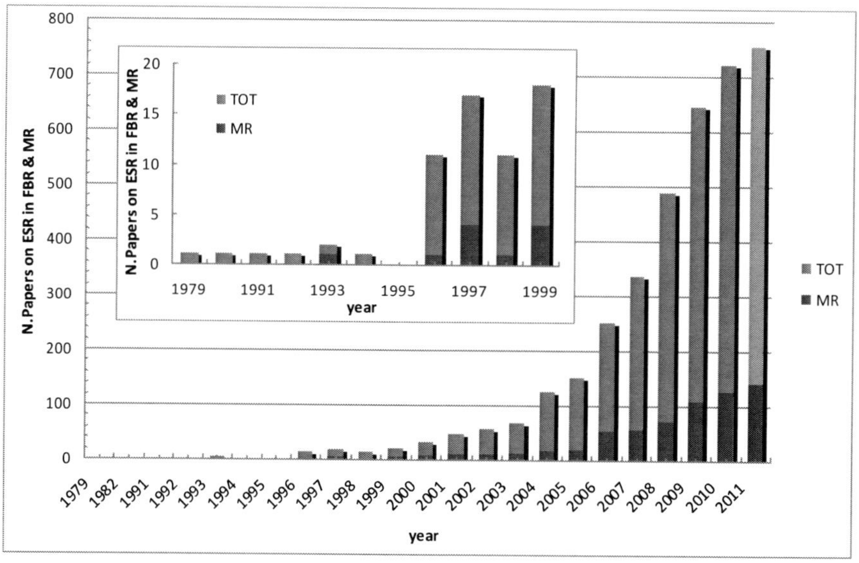

Figure 6. Number of scientific publications vs year on ESR reaction in both FBR and MR.

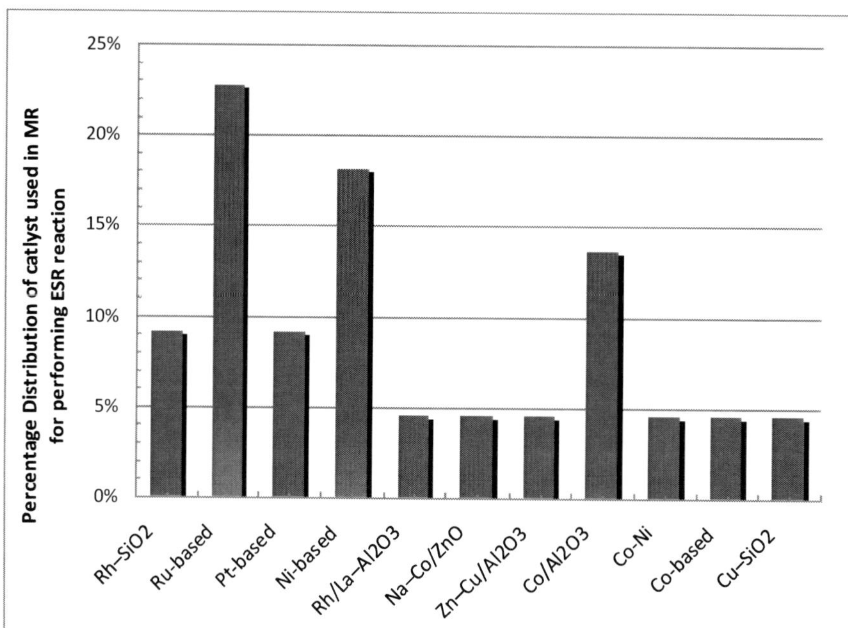

Figure 7. Percentage distribution of the most used catalysts in ethanol reforming reactions performed in MRs.

Table 2 resumes the most significative MR performances of various scientific studies present in the open literature. The experimental data below reported are qualitatively compared each other with the scope of illustrating the different performances achievable using different types of MRs. In particular, Table 2 shows the MRs operating conditions as well as the type of inorganic membranes housed in the MR. As a general consideration, it appears quite evident that Pd-based membranes are the most used both in terms of laminated and supported.

Table 2. Performances and operating conditions of several MRs for ethanol reforming processes present in the open literature

Membrane type	H_2O/ EtOH	O_2/ EtOH	T [°C]	p [bar]	Catalyst	H_2 recovery [%]	H_2 yield [%]	Ref.
dense self-supported Pd-Ag	13/1	-	450	2.0	Pt-Al_2O_3	-	60	[55]
dense self-supported Pd-Ag	18.7/1	-	400	1.5	Co-Al_2O_3	30	18[1]	[56]
dense self-supported Pd-Ag	18.7/1	-	400	3.0	Co-Al_2O_3	90	53[1]	[57]
dense self-supported Pd-Ag	3/1	-	400	1.3	Ru-Al_2O_3	22	56[3]	[58]
dense self-supported Pd-Ag	11/1	-	500	3.6	Ru-Al_2O_3	25	-	[59]
supported Pd-Ag	12/1	-	700	69.0	Rh-$LaAl_2O_3$	-	65	[60]
Pd-Ag supported onto PSS	-	-	450	10.0	Cu-Zn/Al_2O_3	80	-	[62]
Pd-Ag supported onto PSS	1/1	0.2/1	450	9.0	-	-	40	[63]
dense self-supported Pd-Ag	11/1	0.6/1	400	2.5	Ru-Al_2O_3	30	18	[64]
laminated Pd-Ag	-	-	350	1.2	Cu-Zn/Al_2O_3	-	40[2]	[65]
silica-alumina composite	3/1	-	350	-	Co-ZnO	-	23	[66]
SiO_2 supported onto PSS	5/1	-	600	-	Rh-SiO_2	85	50	[67]

[1] high purity H_2 yield = H_2 permeate/6·$EtOH_{feed}$.
[2] calculated.
[3] at H_2O/EtOH = 9/1.

Some studies in literature [54-57] focus on ethanol reforming reactions paying special attention, in the majority of them, to the influence of steam to ethanol feed molar ratio and working at relatively low reaction temperature (400 - 450 °C) using both PSS supported Pd-based and dense self supported Pd-Ag membrane reactors. Concerning the self supported Pd-membranes, they are manufactured by cold-rolling and welding technique, 50 μm thick and joined to two stainless steel ends useful for the membrane housing in the MR. Furthermore, this Pd-Ag membrane is full perm-selective to hydrogen permeation. In particular, Iulianelli et al. [56,57] performed ESR reaction at 400 °C catalyzed by Co-Al_2O_3, obtaining a hydrogen yield ranging from ~ 20% to more than 50% as well as a $CO_{x\text{-free}}$ hydrogen recovery (defined as the CO_x-free hydrogen collected in the permeate side on the total hydrogen produced during the reaction) from 30% to around 90%, by varying the reaction pressure between 1.5 – 3.0 bar and keeping constant the permeate pressure at 1.0 bar. The authors pointed out that higher pressures favour the aforementioned "shift effect" owing to an increase of the hydrogen permeation driving force (see Eq. 15). This involves in a greater removal of hydrogen from the reaction side to the permeate side as a $CO_{x\text{-free}}$ hydrogen stream.

Also in other papers [58,59], a self supported Pd-Ag MR was used for producing hydrogen from ESR reaction. In one case [58], around 20% of high purity hydrogen recovery was achieved working at 1.3 bar of reaction pressure, 400 °C and stoichiometric H_2O/EtOH feed ratio. In the other case [59], although the experimental campaign was performed at higher temperature and pressure, both positive for improving the hydrogen permeation through the membrane, only around 25% hydrogen recovery as a CO_x-free stream was reached. This was probably due to the influence of a higher H_2O/EtOH feed ratio (11/1), which diluted the retentate stream lowering the hydrogen partial pressure on this side, globally reducing the hydrogen permeation driving force and, as a consequence, the hydrogen recovery.

Papadias et al. [60,61] investigated ESR reaction at steam to carbon molar ratio varying between 3/1 and 12/1, using a supported Pd-Ag MR packed with a Rh-$LaAl_2O_3$ catalyst. At 700 °C and 6.9 bar, this MR gives around 75% hydrogen yield, while at 69 bar it drops to 42% owing to an increase of methane yield.

A Pd-Ag onto a PSS membrane was used by Lin et al. [62] to perform ESR reaction in the MR. The membrane was prepared by electroless plating and the experimental tests were carried out at 450 °C, in the pressure range of 2.0 - 11.0 bar, using a Cu-Zn/Al_2O_3 catalyst. A sweep-gas was used for improving the flux of permeating hydrogen and both higher temperatures and pressures advantageously affect the hydrogen permeation driving force, favouring greater hydrogen recovery. As reported in Table 2, the best results were reached at 10.0 bar, with a hydrogen recovery of 80%. However, Lin et al. experimented also the effect of oxygen addition during hydrogen production via ESR reaction [63]. As a result, by adding oxygen, CO_2 selectivity increases, while CO selectivity remains almost the same. Hydrogen production is improved at higher pressures owing to the effect of a higher driving force, even though an increased oxygen input slightly lowers the hydrogen production. This happens owing to the fast oxidation reaction of hydrogen, which indirectly reduces the hydrogen permeation through the membrane.

In the same field of oxidative ESR reaction using self supported Pd-Ag MR, also Iulianelli et al. [64] confirmed that, by increasing the oxygen content on the reaction side (O_2 over ethanol feed ratio > 1.3/1), both hydrogen yield and hydrogen recovery are lowered owing to the fast oxidation reaction of the hydrogen produced during the reaction. At O_2 on ethanol feed ratio equal to 0.6/1, 400 °C and 2.5 bar, ~ 18% hydrogen yield and 30% hydrogen recovery were reached.

Oxygen addition to ethanol dehydrogenation reaction was studied also by Amandusson et al. [65] using a laminated 25 μm PdAg membrane foil located within two reaction chambers. They operated the ethanol dehydrogenation at 350 °C and, as above discussed, observed that hydrogen is decreased with increasing the oxygen content in the feed. These authors pointed out that there is a delicate balance in oxygen/ethanol supply. Indeed, on one hand at too low oxygen content the carbonaceous layer produced during the dehydrogenation reaction of ethanol is not fully oxidised and the hydrogen permeation is lowered; on the other hand, at too high oxygen content, the hydrogen oxidation takes place decreasing the hydrogen permeating through the membrane. In detail, the authors observed that when oxygen is not supplied the carbonaceous layer does not allow the hydrogen permeation. At oxygen/ethanol feed ratio > 1, the hydrogen permeation is lowered by the oxidation reaction of hydrogen, while at oxygen/ethanol feed ratio < 0.7, the carbonaceous layer is not fully oxidised but a constant hydrogen permeation (and recovery) can be achieved. However, the best hydrogen yield reached by the authors was around 40%.

Other researchers approached to ESR reaction using non palladium-based MR with the scope of reducing the cost of palladium. As an example, Lim et al. [66] performed this reaction at 350 °C in a silica-alumina composite MR with a water/ethanol feed ratio ranging between 3/1 and 13/1. The membrane was manufactured by chemical vapour deposition technique and it is not fully perm-selective to hydrogen permeation. The membrane shows at 350 °C H_2/CH_4 gas selectivity equal to 350 with hydrogen permeance in the order of 10^{-8} $mol \cdot m^{-2} \cdot s^{-1} \cdot Pa^{-1}$. At water/ethanol feed ratio equal to 3/1, more than 20% of hydrogen yield was achieved. The authors concluded that the yield of hydrogen is improved as 10-20% more than that of a FBR working at the same MR operating conditions, owing to the effect of the membrane on the reaction system.

An important contribute on hydrogen production from ESR reaction via MR technology was given by Yu et al. [67]. They used a non palladium-based MR for collecting hydrogen with very low CO content (molar concentration < 1%). The membrane was synthesized as SiO_2 supported onto PSS. The reaction is conducted on one side of the MR by means of a Ru-SiO_2 catalyst and the stream permeated through the membrane is rich in hydrogen and CO, which is further converted via water gas shift reaction to decrease the CO content under 1% in the permeate stream. As shown in the Table 2, at water/ethanol feed ratio equal to 5/1 and 600 °C, the hydrogen recovery and the hydrogen yield were around 85% and 50%, respectively.

As a general comment on the data summarized in Table 2, it is possible to say that the experimental results may vary greatly depending on the operating conditions as well as on the typology of inorganic membrane utilized. Thus, a direct quantitative comparison among them is not possible, but only from a qualitative point of view.

Another important parameter taken into account by the researchers is the conversion of ethanol describing the effect of the inorganic membrane on the process. Even in this case, it is not possible to compare directly the experimental results of the several studies present in the specialized literature because they are obtained at different operating conditions. Then, Figure 8 summarizes qualitatively most of the MR performances during ethanol reforming reactions present in the open literature in terms of conversion against temperature. Furthermore, some of the most significative results using FBRs packed with the most active catalysts for this reaction (Rh, Ni, Pt, Pd and Co-based) are reported for a qualitative comparison. Firstly, considering that ESR is an endothermic reaction, ethanol conversion for both MRs and FBRs is favoured by a temperature increase. In the case of Pd-based MRs, as higher the temperature as higher the hydrogen permeating flux, allowing the shift of the reaction equilibrium towards the reaction products, globally favouring a further improvement of conversion. However, Figure 8 points out an important aspect: by using FBRs, most of the ethanol conversions higher than 80% are achieved at T > 600 °C. Only in the case of FBRs packed with Co-based catalysts, the conversions overcome 80% at milder temperatures (450-500 °C). Otherwise, high conversions are reached by MR technology in the range of 300 – 400 °C, non depending on the catalyst used during the reaction.

This ability makes possible for the MRs an energy saving over the FBRs. Furthermore, full perm-selective Pd-based MRs seem to be an important choice in this contest because they are able to perform the ESR reaction with complete conversion achieved at middle temperature (350 – 400 °C) with the further advantage of producing a high purity hydrogen stream to supplying, for example, PEM fuel cells.

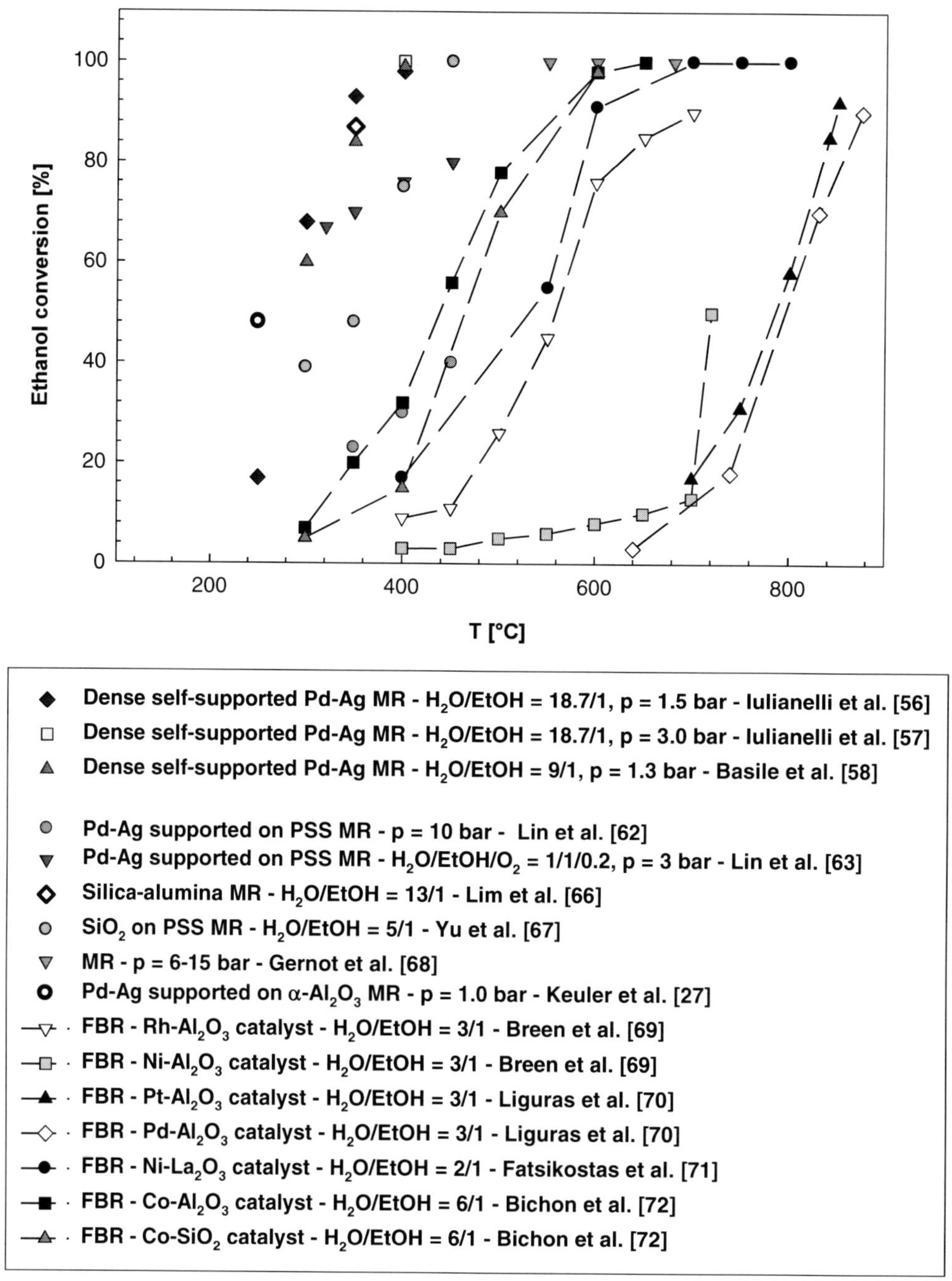

Figure 8. Ethanol conversion vs reaction temperature for several MRs and FBRs studies present in the open literature.

4. Future Perspectives for Ethanol Reforming Reactions in MRs

This chapter showed that MR technology is able to combine two distinct sciences such as catalysis and inorganic membranes. In the scientific community, it is difficult to establish which of them has more relevance in MR development. However, it is reasonable to understand that the exploitation of renewable sources like bio-ethanol could represent a key factor in the viewpoint of hydrogen production via reforming reactions by MR technology, particularly to fulfil the requirements for the integration of fuel processors with PEM fuel cells. One of the issues that is rarely addressed in the open literature concerns this concept: while natural gas (and more in general derived fossil fuels) is essentially utilized for stationary applications, it is expected that alcohols such as ethanol or other liquid fuels such as ammonia will have a major impact in the future portable applications.

A problem still existing in the area of ethanol reforming reactions performed in MRs is the absence of a common protocol to test and to compare the performances (in terms of conversion, hydrogen yield and hydrogen recovery), owing to the different operating conditions used during the experimental campaign. Furthermore, although there is a number of studies on ESR reaction performed in MRs cited in the literature, there is a lack of economical analyses on the applicability of these tools. This could be accounted for the reason that MR technology still shows some deficiencies to overcome before its implementation at larger scales.

Another important issue is the effect of impurities on metallic membranes. Commonly, when using MRs housing membranes hydrogen full perm-selective (i.e. dense Pd and Pd-alloy based membranes), hydrogen permeation may show some deviations from Sieverts' law owing to contaminants effects on the membrane surface. Once these problems will be solved (i.e. synthesis of defect-free, stable and impurity-resistant membranes, no and/or low palladium content membranes development), the benefits due to the MRs utilization instead of FBRs for performing ethanol reforming reactions at industrial scale could become more realistic.

Last but not least, economic evaluations based on the combination of the ethanol as a renewable source and MR technology would be necessary for stimulating improvements on this important and innovative research field.

Conclusion

Hydrogen is a well recognized energy carrier and, therefore, it will achieve a dominant role in future energy systems. Therefore, there is a growing interest in both industrial and scientific communities to develop innovative systems and processes for producing hydrogen in a more technically, environmentally and economically attractive way, based on the idea of sustainable development.

Currently, steam reforming of natural gas and the exploitation of derived fossil fuels are the most used route for hydrogen production. Furthermore, the conventional processes for hydrogen upgrading in refinery applications is carried out by using PSA, membrane, or cryogenic separation processes. Inorganic membranes are excellent candidates for hydrogen purification, particularly when assembled into MRs, able to combine reaction and separation process in a single unit.

In the last years, ethanol as a bio-derived source is becoming an interesting and important candidate for producing hydrogen via reforming reaction and, combined to inorganic MR technology, could match an important result in this area. In this chapter, a short overview on ethanol reforming processes performed in MRs has been also presented, emphasizing the effect of different typology of inorganic membranes on the reaction performances in terms of hydrogen yield, hydrogen recovery and ethanol conversion. At moment, Pd-based MRs seem to be the dominant applications in this area, especially because of the hydrogen perm-selectivity characteristics of Pd-based membranes.

In conclusion, the future perspectives on performing ethanol reforming via inorganic MRs are resumed as reported below:

1. Development of low or no palladium-based membranes useful for ethanol reforming in MRs.
2. Scale-up of ethanol reforming MRs is one of the most important issue.
3. Future researches should be aimed by the improvement of the membrane mechanical resistance during the ethanol reaction processes, at both relatively high reaction temperatures and pressures.
4. Great attention should be paid on evaluating the effective balance between benefits and drawbacks of applying MR technology to ethanol reforming process to produce hydrogen.
5. More attention should be devoted to experimental tests on the lifetime of MRs utilized for carrying out the ethanol reforming processes to validate them as possible alternative to the conventional systems at larger scales.

REFERENCES

[1] A. Heinzel, B. Vogel, P. Hiugner, Reforming of natural gas hydrogen generation for small scale stationary fuel cell systems, *J. Power Sou.*, 105 (2002) 202-207.

[2] P. Moriarty, D Honnery, Intermittent renewable energy: The only future source of hydrogen?, *Int. J. Hydrogen En.*, 32 (2007) 1616-1624.

[3] J. Wee, Applications of proton exchange membrane fuel cell systems, *Renew. Sust. Energy Rev.*, 11 (2007) 1720-1738.

[4] M M. Mulder, *Basic Principles of Membrane Technology*, Kluwer Academic: Dordrecht (1996) ISBN: 0-7923-4247-X (HB).

[5] G. Saracco, V. Specchia, Catalytic inorganic membrane reactors: present experience and future opportunities, *Catal. Rev. Sci. Eng.*, 36 (1994) 305-384.

[6] J. Shu, B.P.A. Grandjean, A. Van Neste, S. Kaliaguine, Catalytic palladium-based membrane reactors: a review, *Can. J. Chem. Eng.*, 69 (1991) 1036-1060.

[7] G.Q. Lu, J.C. Diniz de Costa, M. Duke, S. Giessler, R. Socolow, R.H. Williams, T. Kreutz, Inorganic membranes for hydrogen production and purification: A critical review and perspective, *J. Coll. Interface Sci.*, 314 (2007) 589-603.

[8] T.L. Ward, T. Dao, Model of hydrogen permeation behaviour in palladium membranes, *J. Membrane Sci.*, 153 (1999) 211-231.

[9] G.J. Grashoff, C.E. Pilkington, C.W. Corti, The purification of hydrogen – A review of the technology emphasing, the current status of palladium membrane diffusion", *Plat. Met. Rev.*, 27 (1983) 157-168.

[10] H.P. Hsieh, Inorganic membrane reactors – A review, AIChE Symp. Ser., 85 (1989) 53-67.

[11] F.A. Lewis, K. Kandasamy, B. Baranowski, The "Uphill" diffusion of hydrogen – Strain-gradient-induced effects in palladium alloy membranes, *Plat. Met. Rev.*, 32 (1988) 22-26.

[12] K. Hou, R. Hughes, Preparation of thin and highly stable Pd/Ag composite membranes and simulative analysis of transfer resistance for hydrogen separation, *J. Membrane Sci.*, 214 (2003) 43-55.

[13] G. Maggio, S. Freni, S. Cavallaro, Light alcohols/methane fuelled molten carbonate fuel cells: a comparative study, *J. Power Sou.*, 74 (1998) 17-23.

[14] H. Song, L. Zhang, R.B. Watson, D. Braden, U.S. Ozkan, Investigation of bio-ethanol steam reforming over cobalt-based catalysts, *Catal. Today*, 129 (2007) 346-354.

[15] M. Ni, D.Y.C. Leung, M.K.H. Leung, A review on reforming bio-ethanol for hydrogen production, *Int. J. Hydrogen En.*, 32 (2007) 3238-3247.

[16] M. Pfeffer, W. Wukovits, G. Beckmann, A. Friedl, Analysis and decrease of the energy demand of bioethanol-production by process integration, *Appl. Thermal Eng.*, 27 (2007) 2657-2664.

[17] A. Haryanto, S. Fernando, N. Murali, S. Adhikari, Current status of hydrogen production techniques by steam reforming of ethanol: a review, *En. Fuels*, 19 (2005) 2098-2106.

[18] J.P. Breen, R. Burch, H.M. Coleman, Metal-catalysed steam-reforming of ethanol in the production of hydrogen for fuel cell applications, *Appl. Catal.*, 39 (2002) 65-74.

[19] F. Haga, T. Nakajima, K. Yamashita, S. Mishima, Effect of crystallite size on the catalysis of alumina-supported cobalt catalyst for steam reforming of ethanol, *React. Kinet. Catal. Lett.*, 63 (1998) 253-259.

[20] J. Llorca, N. Homs, J. Sales, P.R. de la Piscina, Efficient production of hydrogen over supported cobalt catalysts from ethanol steam reforming, *J. Catalysis,* 209 (2002) 306-317.

[21] A. Kaddouri, C. Mazzocchia, A study of the influence of the synthesis conditions upon the catalytic properties of Co/SiO_2 or Co/Al_2O_3 catalysts used for ethanol steam reforming, *Catal. Comm.,* 5 (2004) 339-345.

[22] M.S. Batista, R.K.S. Santos, E.M. Assaf, J.M. Assaf, E.A. Ticianelli, Characterization of the activity and stability of supported cobalt catalysts for the steam reforming of ethanol, *J. Power Sou.*, 124 (2003) 99-103.

[23] L. Dolgykh, I. Stolyarchuk, I. Denyega, P. Strizhak, The use of industrial dehydrogenation catalyst for hydrogen production from bioethanol, *Int. J. Hydrogen En.*, 31 (2006) 1607-1610.

[24] [F. Frusteri, S. Freni, V. Chiodo, S. Donato, G. Bonura, S. Cavallaro, Steam and auto-thermal reforming of bio-ethanol over MgO and CeO_2 Ni supported catalysts, *Int. J. Hydrogen En.*, 31 (2006) 2193-2199.

[25] M. Benito, J.L. Sanz, R. Isabel, R. Padilla, R. Arjona, L. Daza, Bio-ethanol steam reforming: Insights on the mechanism for hydrogen production, *J. Power Sou.*, 151 (2005) 11-17.

[26] V. Mas, R. Kipreos, N. Amadeo, M. Laborde, Thermodynamic analysis of ethanol/water system with the stoichiometric method, *Int. J. Hydrogen En.*, 31 (2006) 21-28.

[27] J. N. Keuler, L, Lorenzen, Comparing and modeling the dehydrogenation of ethanol in a plug-flow reactor and a Pd-Ag membrane reactor, *Ind. Eng. Chem. Res.*, 41 (2002) 1960-1966.

[28] K.C. Khulbe, C.Y. Feng, T. Matsuura, Synthetic polymeric membranes, characterization by atomic force microscopy, *Springer* ISBN:3540739939 (2007).

[29] W.J. Koros, Y.H. Ma, T. Shimidzu, Terminology for membranes and membrane processes, *J. Membrane Sci.*, 120 (1996) 149-159.

[30] Y. Xia, Y. Lu, K. Kamata, B. Gates, Y. Yin, Macroporous materials containing three-dimensionally periodic structures, Chemistry of Nanostructured Materials (Ed.: Yang, P.), World Scientific (2003) 69-100.

[31] H.M. Van Veen, M. Bracht, E. Hamoen, P.T. Alderliesten, Feasibility of the application of porous inorganic gas separation membranes in some large-scale chemical processes, Fundamentals of inorganic membrane science and technology, edited A.J. Burggraaf, L. Cot, Elsevier, 14 (1996) 641-681.

[32] K. Briceño, R.G. Valls, D. Montanè, State of the art of carbon molecular sieves supported on tubular ceramics for gas separation applications, *Asia-Pacific J. Chem. Eng.*, 5 (2010) 169-178.

[33] J. Cejka, H. Van Bekkum, A. Corma, F. Schuth, Introduction to zeolite science and practice, Studies in surface science and catalysis, *Elsevier* ISBN: 978-0-444-53063-9 (2007).

[34] A. Julbe, D. Farrusseng, C. Guizard, Porous ceramic membranes for catalytic reactors-overview and new ideas, *J. Membrane Sci.*, 181 (2001) 3-20.

[35] J. Zaman, A. Chakma, Inorganic membrane reactors, *J. Membrane Sci.*, 92 (1994) 1-28.

[36] G. Saracco, H.W.J.P. Neomagus, G.F. Versteeg, W.P.M. van Swaaij, High-temperature membrane reactors: potential and problems, *Chem. Eng. Sci.*, 54 (1999) 1997-2017.

[37] J. Coronas, J. Santamaria, Catalytic reactors based on porous ceramic membranes, *Catal. Today*, 51 (1999) 377-389.

[38] K. Keizer, V.T. Zaspalis, R.S.A. De Lange, M.P. Harold, A.J. Burggraaf, *Membrane reactors for partial oxidation and dehydrogenation reactions, Membr. Proc. Sep. Pur.*, J.G. Crespo and Boddeker, eds., Netherlands, 415-429 (1994).

[39] S. Adhikari, S. Fernand, Hydrogen membrane separation techniques, Ind. *Eng. Chem. Res.*, 45 (2006) 875-881.

[40] R. Dittmeyer, V. Höllein, K. Daub, Membrane reactors for hydrogenation and dehydrogenation processes based on supported palladium, J. *Molecular Catal. A: Chem.*, 173 (2001) 135–184.

[41] A. Basile, Hydrogen production using Pd-based membrane reactors for fuel cells, *Top. Catalysis*, 51 (2008) 107-122.

[42] Y. Zhang, J. Gwak, Y. Murakoshi, T. Ikehara, R. Maeda, C. Nishimura, Hydrogen permeation characteristics of thin Pd membrane prepared by microfabrication technology, *J. Membrane Sci.*, 277 (2006) 203-209.

[43] M.D. Dolan, N.C. Dave, A.Y. Ilyushechkin, L.D. Morpeth, K.G. McLennan, Composition and operation of hydrogen-selective amorphous alloy membranes, *J. Membrane Sci.*, 285 (2006) 30-55.

[44] A. Basile, A. Iulianelli, T. Longo, S. Liguori, M. De Falco, Pd-based Selective Membrane State-of-the-Art, Ch. 2 pp. 21-55 in "Membrane Reactors for Hydrogen Production Processes", L. Marrelli, M. De Falco & G. Iaquaniello Editors, Springer London Dordrecht Heidelberg New York, 2011, ISBN 978-0-85729-150-9, DOI 10.1007/978-0-85729-151-6.

[45] S. Wieland, T. Melin, A. Lamm, Membrane reactors for hydrogen production, *Chem. Eng. Sci.*, 57 (2002) 1571-1576.

[46] A.S. Damle, Hydrogen production by reforming of liquid hydrocarbons in a membrane reactor for portable power generation – Experimental studies, *J. Power Sou.*, 186 (2009) 167-177.

[47] D.J. Edlund, W.A. Pledger, Thermolysis of hydrogen sulfide in a metal-membrane reactor, *J. Membrane Sci.*, 77 (1993) 255-264.

[48] D.J. Edlund, W.A. Pledger, Catalytic platinum-based membrane reactor for removal of H_2S from natural gas streams, *J. Membrane Sci.*, 94 (1994) 111-119.

[49] A. Li, W. Liang, R. Hughes, The effect of carbon monoxide and steam on the hydrogen permeability of a Pd/stainless steel membrane, *J. Membrane Sci.,* 165 (2000) 135-141.

[50] H. Amandusson, L.G. Ekedahl, H. Dannetun, The effect of CO and O_2 on hydrogen permeation through a palladium membrane, *Appl. Surf. Sci.*, 153 (2000) 259-267.

[51] B.A. McCool, Y.S. Lin, Nanostructured thin palladium-silver membranes: Effects of grain size on gas permeation properties, *J. Mater. Sci.,* 36 (2001) 3221-3227.

[52] S. Uemiya, State-of-art-the-art of supported metal membranes for gas separation, *Sep. Pur. Methods*, 28 (1999) 51-85.

[53] D.L. McKinley, W.V. Nitro, *Metal alloy for hydrogen separation and purification*, US patent 3.350.845 (1967).

[54] A. Iulianelli, S. Liguori, V. Calabrò, P. Pinacci, A. Basile, Partial oxidation of ethanol in a membrane reactor for high purity hydrogen production, *Int. J. Hydrogen En.*, 35 (2010) 12626-12634.

[55] A. Basile P. Pinacci, A. Iulianelli, M. Broglia, F. Drago, S. Liguori, T. Longo, V. Calabrò, Ethanol steam reforming reaction in a porous stainless steel supported palladium membrane reactor, *Int. J. Hydrogen En.,* 36 (2011) 2029 -2037.

[56] A. Iulianelli, A. Basile, An experimental study on bio-ethanol steam reforming in a catalytic membrane reactor. Part I: Temperature and sweep-gas flow configuration, *Int. J. Hydrogen En.*, 35 (2010) 3170-3177.

[57] A. Iulianelli, S. Liguori, T. Longo, S. Tosti, P. Pinacci, A. Basile, An experimental study on bio-ethanol steam reforming in a catalytic membrane reactor. Part II: Reaction pressure, sweep factor and WHSV effects, *Int. J. Hydrogen En.*, 35 (2010) 3159-3164.

[58] A. Basile, F. Gallucci, A. Iulianelli, S. Tosti, CO-free hydrogen production by ethanol steam reforming in a Pd-Ag membrane reactor, *Fuel Cells*, 1 (2008) 62-68.

[59] A. Basile, F. Gallucci, A. Iulianelli, M. De Falco, S. Liguori, Hydrogen production by ethanol steam reforming: experimental study of a Pd-Ag membrane reactor and traditional reactor behaviour, *Int. J. Chem. Reactor Eng.*, 6 (2008) A30.

[60] D.D. Papadias, S.H.D. Lee, M. Ferrandon, S. Ahmed, An analytical and experimental investigation of high-pressure catalytic steam reforming of ethanol in a hydrogen selective membrane reactor, *Int. J. Hydrogen En.*, 35 (2010) 2004-2017.

[61] S. Ahmed, S. Lee, D.D. Papadias, Analysis of trace impurities in hydrogen: Enrichment of impurities using a H_2 selective permeation membrane, *Int. J. Hydrogen En.*, 35 (2010) 12480-12490.

[62] W.H. Lin, H-F. Chang, A study of ethanol dehydrogenation reaction in a palladium membrane reactor, *Cat. Today*, 97 (2004) 181-188.

[63] W.H. Lin, C.S. Hsiao, H-F. Chang, Effect of oxygen addition on the hydrogen production from ethanol steam reforming in a Pd-Ag membrane reactor, *J. Membrane Sci.*, 322 (2008) 360-367.

[64] A. Iulianelli, T. Longo, S. Liguori, P.K. Seelam, R.L. Keiski, A. Basile, Oxidative steam reforming of ethanol over Ru-Al_2O_3 catalyst in a dense Pd-Ag membrane reactor to produce hydrogen for PEM fuel cells, *Int. J. Hydrogen En.*, 34 (2009) 8558-8568.

[65] H. Amandusson, L.G. Ekedahl, H. Dannetun, Alcohol dehydrogenation over Pd versus PdAg membranes, *Appl. Catal. A: Gen.*, 217 (2001) 157-164.

[66] H. Lim, Y. Gu, S.T. Oyama, Reaction of primary and secondary products in a membrane reactor: Studies of ethanol steam reforming with a silica-alumina composite membrane, *J. Membrane Sci.*, 351 (2010) 149-159.

[67] C.Y. Yu, D.W. Lee, S.J. Park, K.Y Lee, K.H. Lee, Study on catalytic membrane reactor for hydrogen production from ethanol steam reforming, *Int. J. Hydrogen En.*, 34 (2009) 2947-2954.

[68] E. Gernot, F. Aupretre, A Deschamps, F. Epron, P. Merecot, D. Duprez, C. Etievant, Production of hydrogen from bioethanol in catalytic membrane reactor, 16th Confèrence Mondiale de l'Hydrogène Energie (WHEC16), 13-16 June 2006-Lyon (France), *http://www.ceth.fr/download/presse/art_ceth_3.pdf.*

[69] J.P. Breen, R. Burch, H.M. Coleman, Metal-catalysed steam reforming of ethanol in the production of hydrogen for fuel cell applications, *Appl. Catalysis B: Env.*, 39 (2002) 65–74.

[70] D.K. Liguras, D.I. Kondarides, X.E. Verykios, Production of hydrogen for fuel cells by steam reforming of ethanol over supported noble metal catalysts, *Appl. Catalysis B: Env.*, 43 (2003) 345–354.

[71] A.N. Fatsikostas, X.E. Verykios, Reaction network of steam reforming of ethanol over Ni-based catalysts, *J. Catalysis*, 225 (2004) 439–452.

[72] P. Bichon, G. Haugom, H.J. Venvik, A. Holmen, E.A. Blekkan, Steam reforming of ethanol over supported Co and Ni catalysts, *Top. Catalysis*, 49 (2008) 38-45.

In: Ethanol: Production, Cellular Mechanisms and Health Impact ISBN: 978-1-62257-297-7
Editors: Roy I. Henry and Billy P. Woods

Chapter 4

EPIGENETICS OF FETAL ALCOHOL SPECTRUM DISORDER

Asok Dasmahapatra[1,2,3], Mona Haron[1,2], Doris Powe[1,3,4] and Ikhlas A. Khan[1,5]

[1]National Center for Natural Product Research
[2]Department of Pharmacology, University of Mississippi, Mississippi, US
[3]Jackson State University, Jackson, Mississippi,, US
[4]Tougaloo, Mississippi, US
[5]Department of Pharmacognosy, King Saud University, Saudi Arabia

ABSTRACT

Alcohol is recognized as a classic teratogen capable of inducing a wide range of developmental abnormalities. Exposure to alcohol in utero is the major attributable cause of fetal alcohol spectrum disorder (FASD), which is characterized by craniofacial anomalies and neurodevelopmental impairments, such as cognitive and learning disabilities and behavioral disorders. FASD is recognized as the leading nongenetic cause of mental retardation. Recent studies indicate that epigenetic regulatory mechanisms which include DNA methylation, histone modifications and noncoding RNA expression, play a significant role in FASD development. Aberrant epigenetic alterations are known to disrupt key developmental events, especially the central nervous system. In this review, we examine the potential role of epigenetic factors in the etiology of FASD.

INTRODUCTION

In humans, alcohol exposure during pregnancy has been shown to result in a spectrum of birth defects known as fetal alcohol spectrum disorder (FASD) that can affect a child's growth, development, cognition, and physical appearance (Eaton et al., 2011). FASD is not a diagnostic term like fetal alcohol syndrome (FAS) which generally occurs with chronic consumption of high doses of alcohol during pregnancy and represents the most severe end of the spectrum (Jones et al., 1973). However, FASD, like FAS, has serious CNS, cardiovascular, craniofacial, and skeletal defects that permanently affect an individual both socially and medically (Abel and Hannigan, 1995; Streissguth and O'Malley, 2000; Sokol et al., 2003; Hoyme et al., 2005; Autti-Ramo et al., 2006; Nash et al., 2006; Kodituwakku, 2007; Manning and Eugene, 2007; Spadoni et al., 2007; Thanh and Jonssor, 2009). FASD, the most common preventable cause of mental disability, currently represents the leading cause of mental retardation in North America ahead of Down syndrome and cerebral palsy (Luo, 2009). In most countries, the prevalence of FAS ranges between 0.5 and 2.0 per thousand live births (May and Gossage, 2001), but in some countries it is strikingly higher. In South Africa, for example, 65.2 per 1000 children of school age in Western Cape and 67.2 per 1000 children in Northern Cape Province were diagnosed with FASD (Viljoen et al., 2005; Urban et al., 2008).

Basically, the influence of alcohol on embryonic or fetal development can be classified as (1) direct, when it crosses the placenta and interferes with fetal cells, or (2) indirect, when it induces biochemical or physiological alterations in the mother that can be detrimental to the fetus. All these effects are related to the alterations in genetic mechanisms which lead to a disrupted development. Because of this, FASD is considered to be a multifactorial neurodevelopmental deficit mediated through gene-environment interactions. Other than genetic mechanisms, several studies indicate that epigenetic mechanisms, such as DNA methylation, histone modification, and microRNA interference, are also involved in FASD. However, the evidence in support of epigenetic hypothesis has been weak (Figure 1). In this review, we will provide a brief introduction to epigenetic mechanisms followed by an update on epigenetic changes that are induced by ethanol during prenatal development.

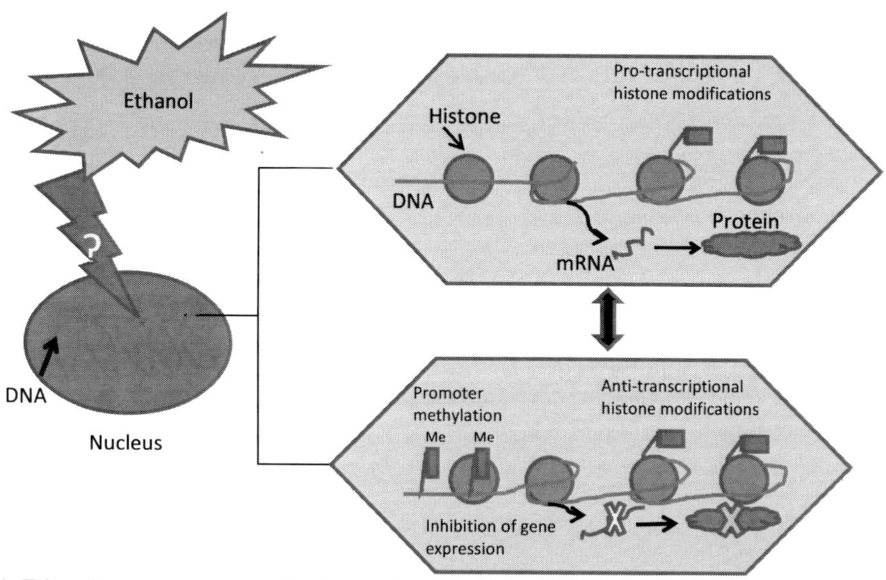

Figure 1. Ethanol can cause changes in chromatin state. The additions of methyl groups to cytosine nucleotides in a regulatory region of a gene inhibit transcription of that particular gene. Histone proteins can also be modified by the addition of various molecules which influences the recruitment and binding of chromatin of various complexes involved in transcription. These histone modifications can either promote or inhibit gene transcription depending upon the nature of modification and the state of the surrounding chromatin.

EPIGENETICS

The literal meaning of the term epigenetics is 'above genetics' which refers to the regulation of gene expression without altering DNA sequences of an organism (Jones and Laird, 1999; Wolffe and Matzke, 1999; Portela and Esteller, 2010, Momparler, 2003; Yoo and Jones 2006). These events modulate chromatin structure, and, in turn, activate or silence gene expression (Lo and Sukumar, 2008; Lutsberg and Ramaswamy, 2009). In contrast to genetic alterations, epigenetic alterations are reversible; moreover, epigenetic marks can be transitory or transmitted transgenerational (Anway et al., 2005; Shukla et al., 2008). During prenatal and early postnatal development, the epigenome is highly susceptible to environmental stimuli such as diet, drugs, environmental agents, and maternal behavior. There are essentially three main stages of development, preconception, preimplantation, and gastrulation, when the generalized global epigenetic remodeling occurs; these stages are highly sensitive to alcohol. As mentioned above, epigenetic marks are mediated through three major mechanisms: DNA methylation, histone medication, and noncoding RNA or miRNA interference.

DNA METHYLATION

DNA methylation is the most accessible epigenetic mark which results in the addition of a methyl group to the carbon-5 (C5) position in the pyrimidine ring of cytosine (5-methylcytosine) located in the context of cytosine-phosphate-guanine (CpG) dinucleotide in the genomic DNA (Dworkin et al., 2009). The reaction is catalyzed by the enzyme DNA methyltransferase (DNMT) with S-adenosyl-methionine (SAM) as the methyl donor (Suzuki and Miyata, 2006). 5-methylcytosine can further be converted into 5-hydroxymethyl-2′-deoxycytidine by the Ten-Elevan-Translocation (TET) protein (Tahiliani et al., 2009; Ito et al., 2010), however, the function of 5-hydroxylmethylation is unclear. It is estimated that 80% of all CpG dinucleotides in mammalian genome are methylated. The remaining unmethylated CpG residues are mostly localized in the promoter regions of constitutively active genes and are known as CpG islands (Esteller, 2007). DNA methylation at the CpG dinucleotide represents an important regulatory component of mammalian genome. Methylated DNA is correlated with transcriptionally inactive genes, where as actively expressed genes are generally hypomethylated (Bird, 1987; Singal and Ginder, 1999). Cytosine methylation also plays important roles in the process of genomic impriting (Tilghman, 1999) and X-chromosome inactivation (Pfeifer et al., 1990). For mammalian development, appropriate DNA methylation is necessary; otherwise, asynchronous development, which is fatal to the fetus, may occur. Methylation can also occur in nonpromoter regions, but the significance is yet unknown. Wu et al. (2010) have shown that methylation of DNA sequences that are beyond the promoters can promote gene expression.

The catalyzing enzyme DNMT has five isoforms, DNMT1, DNMT2, DNMT3A, DNMT3B, and DNMT3L, of which only three, DNMT1, DNMT3A, and DNMT3B, were shown to possess methyltransferase activity (Lo and Sukumar, 2008; Meeran et al., 2010; Veeck and Esteller, 2010). The most abundant DNMT is DNMT1, which is a maintenance methyltransferases, copying methylation marks after DNA replication. De novo DNA methylation is essential for X-inactivation and genomic imprinting, whereas maintenance DNA methylation is required for cell fate decisions, differentiation, and potency (Mudbhary and Sadler, 2011). The DNMT2, the smallest among the DNMTs, does not methylate DNA; it does, however, participate in the recognition of DNA damage, DNA recombination, and mutation repair (Okano et al., 1999). It remains unclear whether DNMT2 plays any role in human disease (Schaefer and Lyko, 2010). DNMT3A and DNMT3B are *de novo* methyltransferase upregulated in aging cells and have similar C-terminal catalytic domain as in DNMT1. Both DNMT3A and 3B can cooperate with DNMT1 for maintenance methylation (Kim et al., 2002; Rhee et al., 2002; Liang et al., 2002; Chen et al., 2003). DNMT3A-deficient mice display organ defects and lethality within weeks after birth with only minor DNA methylation changes (Okano et al., 1999). Mutations in DNMT3B are associated with immunodeficiency, centromeric instability, and facial anomalies syndrome in humans (Hansen et al., 1999; Ueda et al., 2006). Individual ablation of the DNMT genes such as DNMT1, DNMT3A or DNMT3B leads to a disruption of murine embryonic development (Okano et al., 1999: Wu et al., 2010). Although DNMT3A and 3B have overlapping de novo methylation functions, DNMT3A is critical for imprinting and germ cells, whereas DNMT3B is more essential for early somatic development. DNMT3L is required for the catalytic activity of both DNMT3A and DNMT3B. The amino acid identity between DNMT3A and DNMT3L is very high, but the latter (DNMT3L) lacks any catalytic activity due to the absence of conserved catalytic residues. DNMT3L knockout mice are viable, but sterile (Boure'his and Bestor, 2004).

The mechanisms by which DNMT methylates cytosine C5 have been studied extensively (Wu and Santi, 1987; Chen et al., 1991; O'Gara et al., 1991; Suzuki and Miyata, 2006; Medina-Fanco and Caulfield, 2011). DNMTs can also silence genes other than promoter DNA methylation. The formation of a repressive transcription complex by histone deacetylases (HDACs) and the methyl-CpG binding domain (MBD) family of proteins at the promoter regions did not require promoter methylation (Mistry et al., 2003; Lustberg and Ramaswamy, 2009). Moreover, DNMT1, through its N-terminus, binds to HDAC2 and a DNMT1-associated protein called DMAP1 to form a complex at replication fork (Rountree et al., 2000). This complex converts acetylated histones to the deacetylated inactive form and highlights the interplay between methylation and acetylation processes in epigenetic regulation (Lustberg and Ramaswamy, 2009).

One of the early studies demonstrated that acute alcohol exposure in pregnant mice (gestation days 9-11) resulted in reduced methylation of fetal DNA and DNMT levels (Garro et al., 1991). Another study by Bielawski et al (2002) reported that expression of DNMT1 was reduced in the sperm of paternal rats after nine weeks of ethanol exposure. Analysis of methylation patterns of sperm DNA from human volunteers showed a correlation between chronic alcohol use and DNA demethylation of the regions which are normally hypermethylated (Ouko et al., 2009). These studies suggest that the transmission of these epigenetic changes to the offspring during fertilization may affect prenatal development. Haycock and Ramsey (2009) studied imprinting of the *H19/Igf2* in preimplantation mouse embryos after maternal ethanol exposure and found severe growth retardation of embryos, however, did not find any epigenetic changes at the *H19* imprinting control region. Liu et al. (2009) reported that exposure of whole mouse culture embryos to a high concentration of ethanol (404 mg/dl which is equivalent to 88 mM) altered DNA methylation of genes located on chromosome 7, 10 and X, known to have a role in the cell cycle, growth and apoptosis, with increased methylation in some genes and decreased methylation in others. Substantial evidence indicates that DNA methylation plays a significant role in neural cell linage differentiation and early brain development (Feng et al., 2007; MacDonald and Roskams, 2009; Setoguchi et al., 2006; Takizawa et al., 2001). Alcohol exposure prevented methylation of cut-like 2 (*cult2*), insulin-like growth factor 1 (*Igf1,*), epidermal growth factor-containing fibulin-like extracellular matrix protein (*Efemp1*), and SRY-box-containing gene 7 (*sox7*) which are associated with neural development. Several other genes such as lens intrinsic membrane protein 2 (*Lim2*), which played a significant role in eye development, the epigenetic mark SW1/SNF related, matrix associated, actin dependent regulator of chromatin, subfamily a, member 2 (*Smarca2*), and developmental disorder gene DiGeorge syndrome critical region gene (*Dgcr2*) have an altered methylation pattern after ethanol exposure ((Zhou et al., 2011). Kaminen-Ahola et al (2010) found that in *Agouti* mouse maternal ethanol consumption either before or after fertilization affects the expression of A^{vy} (a dominant mutation of the *Agouti* locus which is epigenetically sensitive) in the offspring having FASD phenotypes. In medaka (*Oryzias latipes*), the promoter region of *Aldh1A2* locus consists 31 CpG dinucleotides of which none are methylated (Hu et al., 2009). Embryonic ethanol exposure (100 and 400 mM of ethanol for 48 h) was unable to alter *Aldh1A2* promoter methylation pattern in embryos or in the tissues of adults developmentally exposed to ethanol (Hu et al., 2009).

DNA methylation depends on dietary methionine and folate (Figure 2). Diets deficient in methionine have been reported to lead to DNA hypomethylation (Wilson et al., 1984; Kobor and Weinberg, 2011). Moreover, alcohol has multiple effects on methyl donors (Mason and Choi, 2005) and probably interferes with the folate-methyl metabolic pathway for methyl donors by inhibiting methionine synthase and methionine adenosyltransferase. Alcohol also prevents folic acid absorption at the systemic and cellular level (McGeer et al., 1983; Hamid et al., 2009). Inhibition of methionine synthase creates a "methylfolate trap" which is analogous to that which occurs in vitamin B12 deficiency (Cravo and Camilo, 2000; Mason and Choi, 2005). In animals, alcohol ingestion inhibited folate-mediated methionine synthesis and interrupted critical methylation processes mediated-through SAM. In addition, other evidence indicates that alcohol may redirect the utilization of folate toward serine synthesis and, thereby, interfere with a critical function of methylene tetrahydrofolate and methionine-homosystein cycle by disrupting the enzymes required for methionine metabolism (Barak et al., 1993; Lieber 1988; Lu et al., 2000). Choline, a key component of the methionine-homocysteine cycle, is also a target of alcohol's teratogenic effects (Zeisel, 2011).

The role of DNA methylation in ethanol-related changes in NMDA receptor (NR2B subunit) was also investigated (Marutha Ravindran and Ticku, 2004, 2005). Chronic ethanol treatment was shown to upregulate *NR2B* gene expression; however, NR2B levels were similar to control levels after 48h of ethanol withdrawal (Kalluri et al., 1998; Marutha Ravindran et al., 2005; Rani et al., 2005). Furthermore, it was found that upregulation of NR2B subunit expression correlated to a reduction in CpG island methylation after chronic ethanol exposure (Marutha Ravindran et al., 2005). Acute ethanol treatment, however, neither affected DNA methylation nor changed the *NR2B* gene expression (Marutha Ravindran et al., 2005). Studies in human showed that acute alcohol intake increased DNA methylation in the promoter region of the α-synuclein gene (Bonsch et al., 2005). Thus alcohol exposure through downstream epigenetic modification may bridge the gene environment interaction leading to development of neurodevelopmental disorders such as FASD.

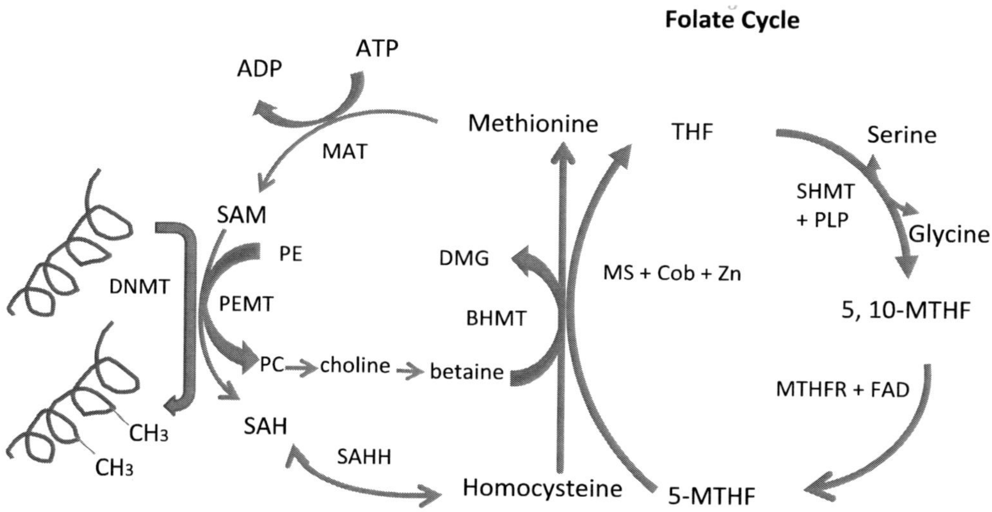

BHMT= betaine:homocysteine methyltransferase; Cob= cobalamine; DHF= dihydrofolate; DMG= dimethyl glycine; FAD= flavin adenine dinucleotide; MAT= Methionine adenosyltransferase; MTHFR=methylenetetrahydrofolate reductase; MS= methionine synthase; MTRR= Methionine synthase reductase; 5-MTHF= 5-methyltetrahydrofolate; 5, 10-MTHF= 5, 10-methyllenetetra hydro folate; PLP= pyridoxal phosphate; SAH= S-adenosylhomocysteine; SAHH= SAH hydrolase: SAM= S-adenosylmethionine; THF=tetrahydrofolate.

Figure 2. Methionine-homocysteine cycle which generates methionine, a precursor of S-adenosyl methionine (SAM) which serves as the methyl donor for methylation reactions. DNA methyltransferases (DNMTs) catalyze the transfer of methyl groups from SAM to DNA or proteins which result an altered function of DNA or protein molecules. The demethylated SAM further processed to homocysteine, which in turn converted to back to methionine through folate-dependent or –independent mechanisms.

Histone Modification

Another prevalent epigenetic mechanism influencing gene transcription is histone modification. Many evidences of the crosstalk between DNA methylation and histone modifications have also been reported (Jones, 2002; Kouzarides, 2007; Vaissiere et al., 2008). Histone proteins are responsible for maintaining chromatin structure either in an accessible or inaccessible state to various transcriptional activators and repressors for their binding to gene promoters (Dhalluin et al., 1999). The euchromatic DNA, which contain four core histones (H2A, H2B, H3 and H4), form an octamer around which 147 bp of DNA are wrapped forming a nucleosome, the basic building block of chromatin structure (Luger et al., 1997; Ito and Adcock, 2002; Strahl and Allis, 2000; Davis and Ross, 2007; Minard et al., 2009). Histone H1 acts as a linker between the nucleosomes (Shukla et al., 2008). The loosened state of the chromatins is referred to as euchromatin which is associated with active gene transcription. However, the condensed state of the chromatins known as heterochromatin in which the genes are thought to remain in a silenced state. Histone modifications involve a number of different post-translational modifications to the lysin rich tail regions of histones. These modifications include acetylation at lysine, methylation at lysine and arginine, phosphorylation at serine and threonine, sumoylation and ubiquitiniaiton at lysine, ADP ribosylation, deamination, proline isomerization, and propionylation (Schubeler et al., 2004; Shilatifard, 2006; Liu et al., 2009). Heterochromatin state is associated with methylation of CpG dinucleotides, hypoacetylation of histones H3 and H4, and demethylation/trimethylation of lysine 9 on histone H3. In contrast to heterochromatin, euchromatin is associated with hypomethylation of CpG dinucleotides, acetylation of histones H3 and H4 and demethylation/trimethylation of lysine 4 on histone H3 (Arney and Fisher, 2004). All of these modifications can alter the accessibility of DNA to transcriptional activity by two ways: (1) affecting electrostatic links between DNA and histone proteins which cause chromatin loosening and thereby enhancing access of transcriptional machinery to the DNA, and (2) by changing histone-histone interaction which impacts chromatin reassembly following transcription. Each modification targets different amino acid residues. Most of these modifications are reversible (Davis and Ross, 2007). The most important histone modifications, having effects on gene expression, are located on histone H3 and histone H4 (Bronner et al., 2010]. One, that is known to have a gene silencing role and to have a strong relationship with DNA methylation, is the di or tri-methylation of lysine 9 of histone 3 (H3K9me2 or H3K9me3). But methylation on the same histone on lysine 4 (H3K4me) is related to gene activation (Peterson and Laniel, 2004; Sims and Reinberg, 2006). The histone modifications are catalyzed by specific enzymes (Lo and Sukumar, 2008).

Acetylation of histones occurs within the amino-terminal tails protruding from the surface of the nucleosomes and also on the globular core region by transfer of acetyl group from acetyl-Co-A to ε amino group of lysine residues (N^{ε}). The reaction is catalyzed by histone acetyl transferase (HAT). The major sites of acetylation of histone H3 are lys 4, lys 9, lys 14 and lys 28. The global acetylation of histone tails decreases electrostatic interactions between the negatively charged DNA and the basic lysine residues (Travers and Thompson, 2004). As a result, the chromatin becomes more open to provide accesses for transcription factors to DNA. Conversely, deacetylation which is catalyzed by histone deacetylases (HDACs) removes the neutralizing acetyl charge and induces chromatin condensation and gene inactivation and silencing (Kouzarides, et al., 2007). HATs also acetylate several non-histone proteins such as GATA1, E2F1, pRB, or TP53 frequently associated with cellular transformation (Martinez-Balbas et al., 2000). HATs cannot bind to target gene promoters directly; however, they are recruited by DNA-bound transcription factors (Roth et al., 2001). But HDACs, whose primary role is to oppose the activity of HATs, are recruited to the hypermethylated CpG islands of tumor suppressor genes via MBDs or independent of methylation by specific transcription factors (Deckert and Struhl 2001). Many nonhistone proteins are also used as substrates for HDACs (Deckert and Struhl, 2001; Gray and Ekstrom, 2001; Mclaughlin and La Thangue, 2004; Robertson et al., 2000). HATs can be classified into five families: the GNAT family, the MYST family, the p300/CBP family, the SRC family, and the TAFII250 family (Roth et al, 2001; Suzuki and Miyata, 2006). Until now, eighteen human HDACs have been identified and are grouped into four classes: Class I, II, III and IV, based on their homology with yeast HDACs and their phylogenetic conservation (Gregoretti et al., 2004). Class I HDACs (HDAC1, 2, 3 and 8) share sequence similarity with the yeast RPD3 deacetylase, are ubiquitously expressed, and are localized mainly in the nucleus. Class II HDACs (HDAC4, 5, 6, 7, 9 and 10) are homologous to the yeast Hda1 deacetylase, are both nuclear and cytoplasmic, and are restricted to certain tissue. Class II HDACs are further subdivided into class IIa (HDAC4, 5, 7 and 9) and class IIb (HDAC6 and 10). Class III HDACs are represented by sirtuins (SIRT1 to SIRT7), a family of seven HDACs sharing homology with yeast silent information regulator 2 (Sir2). Class IV has only one member, HDAC11, which shares conserved residues with both class I and II HDACs (Marks and Xu, 2009). Class I, II and III HDACs have been implicated in the DNA damage response, homolologous recombination (HR), and chromatin integrity.

Histone methylation occurs in lysine and arginine residues, catalyzed by histone lysine methyl transferases (lysine) or histone arginine methyl transferases (arginine) (Yamane et al., 2007) where the methyl group is donated by SAM. The major H3 histone methylation sites at lysine are lys4, lys9, lys27, lys36, lys79, and on H4 is Lys20. At present, there are twenty-four known sites of methylation of histones: seventeen are lysine residues and seven are arginine residues (Bannister and Kouzarides, 2005). Furthermore, lysine side chains may be mono-, di-, or trimethylated, whereas the arginine side chain may be either mono- or dimethylated. In contrast to histone acetylation, histone methylation does not change histone's charge (Dalvai and Bystricky, 2010). Histone methylation can be repressive or activating depending on the types of lysine residues and the degree of methylation (Mudbhary and Sadler, 2011). Recruitment of specific motif bearing factors to the methylated lysine is responsible for affecting chromatin structure. Histone demethylation is catalyzed by histone demethylases (Shi et al., 2004; Tsukada, et al., 2006).

Selective acetylation of histone H3 at lys 9 by ethanol was observed in primary culture of rat hepatocytes (Park et al., 2003), and in hepatic stellate cells (Kim and Shukla, 2006). Histone acetylation was also observed in lung, spleen, and testes of rats treated with ethanol in vivo (Kim and Shukla, 2006). Involvement of oxidative stress in the induction of H3 histone acetylation by alcohol was observed in the primary hepatocyte culture of rats (Choudhury et al., 2010). Chronic ethanol feeding to rats through liquid diets for one to four weeks was unable to increase in global acetylation of H3-Lys9, however, was able to increase the association of H3-Lys9 in the alcohol dehydrogenase I (ADH1) gene (both in promoter and in the coding region). In contrast, the association of acetylated H3-lys9 with iNOS and c-jun genes was not found (Park et al., 2012). In the nervous system, the best characterized histone modification is acetylation of N-terminal lysine residues, which generally shifts the confirmation of chromatin into a relaxed state that leads to an upregulation in gene expression (Renthal and Nestler, 2009). Pandey et al (2008) reported that acute administration of an anxiolytic dose of ethanol (1 g/kg intraperitoneally; blood alcohol level=92 mg/dl) decreased acetylated H3/H4 in the central and medial nuclei of the amygdala with decreased HDAC activity. However, rats chronically treated with an ethanol-containing liquid diet for fifteen to sixteen days (Blood alcohol level=177 mg/dl) and then withdrawn for twenty-four hours decreased the levels of acetylated H3/H4, but HDAC activity increased in amygdala. In the pathophysiology of FASD, alteration in histone modification remains an open question. Neonatal rats (postnatal days two to twelve which are considered as in third trimester of human pregnancy) exposed to ethanol vapor four hours per day have shown reduced acetylated H3/H4 levels at postnatal days two to ten without altering the H3/H4 protein levels in cerebellum (Guo et al., 2011). Other mechanisms of histone modification remained almost unexplored in FASD.

MicroRNA

The most recently emerged participant in the epigenetic field is a family of small regulatory RNAs called microRNAs (miRNA) which are small non-coding RNAs ranging in size from nineteen to twenty-four nucleotides that inhibit protein expression of target genes (Lewis et al., 2005). These RNAs regulate gene expression by sequence-specific binding to the 3′-untranslated region (3′UTR) of target mRNAs, resulting in either mRNA degradation or inhibition of mRNA translation (He and Hannon, 2004). It is estimated that in higher eukaryotic genomes, miRNAs account for ~1% of predicted genes,and, in humans, up to 30% of genes might be regulated by miRNAs (Griffiths-Jones, 2004). Current bioinformatics tools predict that each miRNA recognizes an average of 100-200 different mRNA targets (Lewis et al., 2005). Approximately 1100 human miRNAs are registered in the miRNA data base (Sato et al., 2011). Epigenetic silencing of a miRNA may be a reflection of tissue specificity. Subgroups of miRNAs which are regulated epigenetically have been reported (Sato et al., 2011); moreover, a variety of miRNAs regulate epigenetic pathway-related molecules such as polycomb group proteins, HDACs and DNMTs are also reported (Sato et al., 2011). Extensive analysis of genomic sequences of miRNA genes has shown that approximately half of them (miRNA) are associated with CpG islands (Weber et al., 2007; Lehmann et al., 2008). A line of evidence indicates that miRNAs are involved in the regulation of a number of genes that are involved in development (Karp and Ambros, 2005), cell proliferation (Cheng et al., 2005), apoptosis (Xu et al., 2004), and carcinogenesis (He et al., 2005a, 2005b). Recent studies provide growing evidence for the involvement of miRNA in embryo development and maintenance of tissue identity (Ambros and Chen, 2007; Yu et al., 2007). In Dicer I (RNaseIII enzyme) mutant mice, miRNAs have wide-ranging roles in embryogenesis (Bernstein et al., 2003).

Depending upon the concentration and tissue type, miRNA expression is affected differentially by alcohol. High levels of ethanol (70 mM) which was achieved in alcohol addiction, suppress miR-21, miR-335, miR-9 and miR-153; however, low levels (13 mM) upregulate miR-335 (Sathyan et al. 2007). In adult neurons, upregulation of miR-9 was observed (Pietrzykowski et al., 2008). It was suggested that disturbance of miRNA functions may contribute to the alcohol-induced developmental deficiencies (Wang et al., 2008). Prenatal ethanol exposure upregulated miR-10a, miR-10b, miR-9, miR145, miR-30a-3p and miR152 in fetal brains of C57BL/6J mice, whereas miR-200a, miR-496, mir-296, miR30e-5p, mir-362, miR-339, miR-29c and miR-154 were down regulated (Wang et al., 2009). Both miR10a and mir-10b are embedded in the Hox cluster during evolution, and miR10a, miR-10b and Hox genes are expressed in early embryo development (Wang et al., 2009). In nonmammalian vertebrates, such as zebrafish, embryos exposed to 1% and 1.5% ethanol from 4 hour post fertilization (hpf) to 24 hpf showed upregulation of miR-153a, miR-725, miR-30d, let-7k, miR-100, miR-738 and miR-732. These miRNAs are involved in cell cycle control, apoptosis and transcription in zebrafish development (Soares et al., 2012) in the pathophysiology of FASD.

FUTURE DIRECTIONS

Epigenetic marks are complex, interdependent, and ambiguous. Although some of the recent studies mentioned in this review have pointed out the important roles of epigenetic mechanisms in FASD, the data are consequences of epigenetic remodeling on FASD, and the establishment of a cause and effect relationship, would require additional biochemical and proteomic studies. ` What is clear is that FASD results in lifelong disability, and that there is no cure. As the epigenetic marks, specifically DNA or histone modification, are not coded for in the DNA sequence, vary among cell types, and are mediated in large part by enzymes, epigenetic studies are attractive targets for the development of drugs (epigenetic drugs) for FASD therapies.

REFERENCES

Abel EL, Hannigan JH. 'J-shaped' relationship between drinking during pregnancy and birth weight: reanalysis of prospective epidemiological data, Alcohol *Alcohol*, 30:345-355 (1995)

Ambros V, Chen X. *The regulation of genes and genomes by small RNAs. Development,* 134:1635-1641(2007).

Anway MD, Cupp AS, Uzumcu M, Skinner MK. Epigenetic transgenerational actions of endocrine disruptors and male fertility. *Science*, 308:1466-1469(2005).

Arney KL, Fisher AG. Epigenetic aspects of differentiation. *J Cell Sci*. 117:4355-4363(2004).

Autti-Ramo I, Fagerlund A, Ervalahti N, Loimu L, Korkman M, Hoyme HE. Fetal alcohol spectrum disorders in Finland: clinical delineation of 77 older children and adolescents, *Am. J. Med. Genet. A*. 140:137-143 (2006).

Banister AJ, Kouzarides T. Reversing histone methylation. *Nature*, 436: 1103-1106 (2005).

Barak AJ, Beckenhauer HC, Junnila M, Tuma DJ. Dietary beatine promotes generation of hepatic S-adenosylmethionine and protects the liver from ethanol-induced fatty infiltration. *Alcohol Clin Exp Res*. 17:552-555(1993)

Bernstein E, Kim SY, Carmell MA, Murchison EP, Alcorn H, Li MZ, Mills AA, Elledge SJ, Anderson KV, Hannon GJ. Dicer is essential for mouse development. *Nat Genet*. 35:215-217(2003).

Bielawski DM, Zaher FM, Svinarich DM, Abel EL. Paternal alcohol exposure affects sperm cytosine methyltransferase messenger RNA levels. *Alcohol. Clin. Exp. Res*. 26:347-351(2002).

Bird AP. CpG islands as gene markers in the vertebrate nucleus. *Trends Genet*. 3:342-347(1987).

Bonsch D, Lenz B, Kornhuber J, Bleich S. DNA hypermethylation of the alpha synuclein promoter in patients with alcoholism. *Neuroreport* 16:167-170(2005).

Boure'his D, Bestor TH. Meiotic catastrophe and retransposon reactivation in male germ cells lacking Dnmt3L. *Nature*, 431:96-99(2004).

Bronner C, Fuhrmann G, Chedin FL, Macaluso M, Dhe-Paganon S. UHRF1 links the histone code and DNA methylation to ensure faithful epigenetic memory inheritance. *Genet Epigenet*. 2009:29-36(2010).

Chen L, MacMillan AM, Chang W, Ezaz-Nikpay K, Lane WS, Verdine GL. Direct identification of the active-site nucleophile in a DNA (cytosine-5)-methyltransferase. *Biochemistry*, 30:11018-11025 (1991).

Chen T, Ueda Y, Dodge JE, Wang Z, Li E. Establishment and maintenance of genomic methylation patterns in mouse embryonic stem cells by Dnmt3a and Dnmt3b. *Mol cell boil.* 23:5594-5605(2003).

Cheng CM, Byrom MW, Shelton J, Ford LP. Antisense inhibition of human miRNAs and indications for an involvement of miRNA in cell growth and apoptosis. *Nucleic Acids Res.* 33:1290-1297(2005).

Choudhury M, Park PH, Jackson D, and Shukla SD. Evidence for the role of oxidative stress in the acetylation of histone H3 by ethanol in rat hepatocytes. *Alcohol*, 44:531-540(2010).

Cravo ML, Camilo ME. Hyperhomocysteinemia in chronic alcoholism: relations to folic acid and vitamins B(6) and B912) status. *Nutrition*, 16:296-302/

Dalvai M, Bystricky K. The role of histone modifications and variants in regulating gene expression in breast cancer. *J. Mammary Gland Biol. Neoplasia,* 15:19-33 (2010).

Davis CD, Ross SA. Dietary components impact histone modifications and cancer risk. *Nutr Rev.* 65:88-94(2007).

Deckert J, Struhl K. Histone acetylation at promoters is differentially affected by specific activators and repressors. *Mol. Cell. Biol.* 21, 2726-2735 (2001)

Dhalluin C, Carlson JE, Zeng L, He C, Aggarwal AK, Zhou MM. Structure and ligand of a histone acetyltransferase bromodomain. *Nature*, 399: 491-496(1999)

Dworkin AM, Huang TH, Toland AE. Epigenetic alterations in the breast: implications for breast cancer detection, prognosis and treatment. *Semin Cancer Biol.* 19:165-171(2009).

Eaton B, Gangluff D, Mengel M. Fetal alcohol spectrum disorders: flying under the radar, *J Ark Med Soc.* 107:260-262(2011).

Esteller M. Cancer epigenomics: DNA methylomes and histone-modification maps. *Nat Rev Genet.* 8:286-298(2007).

Feng J, Fouse S, Fan G. Epigenetic regulation of neural gene expression and neuronal function. *Pediatr Res.* 61:58R-63R(2007).

Fernandez, S.V., 2011. Estrogen, alcohol consumption, and breast cancer. Alcohol. *Clin. Exp. Res.* 35, 389-391.

Garro AJ, McBeth DL, Lima V, Lieber CS. Ethanol consumption inhibits fetal DNA methylation in mice: implications for the fetal alcohol syndrome. *Alcohol Clin Exp Res.* 15:395-398(1991).

Gray SG, Ekstrom TJ. The human histone deacetylases family. *Exp Cell Res.* 262:75-83(2001).

Gregoretti IV, Lee YM, Goodson HV. Molecular evolution of the histone deacetylases family: functional implications of phylogenetic analysis. *J. Mol. Biol., 338*, 17-31 (2004).

Griffiths-Jones S. The microRNA registry. *Nucleic Acids Res.* 32:D109-D111(2004).

Guo W, Crossey EL, Zhang L, Zucca S, George OL, Valenzuela F, Zhao X. Alcohol exposure decreases CREB binding protein expression and histone acetylation in the developing cerebellum. *PLoS one*, 6:e19351(2011).

Hamid A, Wani NA, Kaur J. New perspectives on folate transport in relation to alcoholism-induced folate malabsorption-association with epigenome stability and cancer development. *FEBS J.* 276:2175-2191(2009).

Hansen RS, Wijmenga C, Luo P, Stanek AM, Canfield TK, Weemaes CM, Gartler SM. The DNMT 3B DNA methyltransferase gene is mutated in the ICF immunodeficiency syndrome. *Proc Natl Acad Sci USA*. 96:14412-14417(1999).

Haycock P, Ramsay M. Exposure of mouse embryos to ethanol during preimplantation development effect on DNA methylation in the H19 imprinting control region. *Biol Reprod.* 81:618-627(2009).

He H, Jazdzewski K, Li W., Liyanarachchi S, Nagy R, Volinia S, Calin GA, Liu CG, Franssila K, Suster S, Kloos RT, Croce CM, de la Chapelle A. The role of microRNA genes in papillary thyroid carcinoma. *Proc Natl Acad Sci USA*. 102:19075-19080(2005).

He L, Hannon GJ. MicroRNAs: small RNAs with a big role in gene regulation. *Nat. Rev. Genet.* 5, 522-531 (2004).

He L, Thompson JM, Hemann MT, Hernando-Monge E, Mu D, Goodson S., Powers S, Cordon-Cardo C, Lowe SW, Hannon GJ, Hammond SM. A microRNA polycistron as a potential human oncogene. *Nature*, 435(7043):828-833.

Hoyme HE, May PA, Kalberg WO, Kodituwakku P, Gossage JP, Truiilo PM, Buckley DG, Miller JH, Aragon AS, Khaole N, Viljoen DL, Jones KL, Robinson LK. A practical clinical approach to diagnosis of fetal spectrum disorders: clarification of the 1986 institute medicine criteria. *Pediatrics*:39-47(2005).

Hu Y, Willett KL, Khan IA, Sceffler BE, Dasmahapatra AK. Ethanol disrupts chondrification of the neurocranial cartilages in medaka embryos without affecting aldehyde dehydrogenase 1A2 (Aldh1A2) promoter methylation. *Comp. Biochem. Physiol.* 150C, 495-502(2009).

Ito K, Adcock IM. Histone acetylation and histone deacetylation. *Mol. Biotechnol.* 20, 99-106 (2002).

Ito S, D'Alessio AC, Taranova OV, Hong K, Sowers LC, Zhang Y.Role of tet proteins in 5mc to 5hmc conversion, ES-cell self-renewal and inner cell mass specification. *Nature*, 466(7310): 1129-1133(2010).

Jones PA. DNA methylation and cancer. *Oncogene*, 21: 5358-5360 (2002).

Jones PA, Laird PW. Cancer epigenetics comes of age. *Nat Genet*. 21:163-167(1999)

Jones KL, Smith DW, Ulleland CN, Streissguth AP. Pattern of malformation in offspring of chronic alcoholic mothers. *Lancet*, 301(7815):1267-1271(1973).

Kalluri HS, Mehta AK, Ticku MK. Up-regulation of NMDA receptor subunits in rat brain following chronic ethanol treatment. *Brain res Mol Brain Res*. 58:221-224(1998).

Kaminen-Ahola N, Ahola A, Maga M, Mallitt KA, Fahey P, Cox TC, Whitelaw E, Chong S. Maternal ethanol consumption alters the epigenotype and the phenotype of offspring in a mouse model. *PLoS Genet.* 15: e1000811 (2010).

Karp X, Ambros V. Developmental biology. Encountering microRNAs in cell fate signaling. *Science*, 310:1288-1289(2005).

Kim GD, Ni J, Kelesoglu N, Roberts RJ, Pradhan S. Cooperation and communication between the human maintenance and de novo DNA (cytosine-5) methyltransferases. *EMBO J.* 21:4183-4195(2002).

Kim JS, Shukla SD. Histone H3 modifications in rat hepatic stellate cells by ethanol. *Alcohol Alcohol*. 40:367-372(2005).

Kim JS, Shukla SD. Acute in vivo effect of ethanol (binge drinking) on histone H3 modifications in rat tissues. *Alcohol Alcohol* 41:126-132(2006).

Kobor MS, Weinberg J. Focus on: Epigenetics and fetal alcohol spectrum disorders. *Alcohol Res Health*, 34: 29-37 (2011).

Kodituwakku PW. Defining the behavioral phenotype in children with fetal alcohol spectrum disorders: a review. *Neurosci. Biobehav. Rev*. 31:92-201 (2006).

Kouzarides T. Chromatin modifications and their function. *Cell*, 128:693-705 (2007).

Lehmann U, Hasemeier B, Christgen M, Muller M, Romermann D, Langer F, Kreipe H. Epigenetic activation of microRNA gene hsa-mir-9-1 in human breast cancer. *J. Pathol.* 214, 17-24 (2008).

Liang G, Chan MF, Tomigahara Y, Tsai YC, Gonzales FA, Li E, Laird PW, Jones PA. Cooperativity between DNA methyltransferases in the maintenance methylation of repetitive elements. *Mol cell Biol.* 22:480-491(2002).

Lieber CS. Biochemical and molecular basisof alcohol-induced injury to liver and other tissues. New Eng J Med. 319:1639-1650(1988).

Lewis BP, Burge CB, Bartel DP. Conserved seed pairing, often flanked by adenosines, indicates that thousands of human genes are mocroRNA targets, *Cell*, 120:15-17 (2005).

Liu Y, Balaraman Y, wang G, Nephew KP, Zhou FC. Alcohol exposure alters DNA methylation profiles in mouse embryos at early neurulation. *Epigenetics*, 4:500-511 (2009).

Lo PK, Sukumar S. Epigenetics and breast cancer. *Pharmacogenomics* 9:1879-1902(2008).

Lu SC, Huang ZZ, Yang H, Mato JM, Avila MA, Tsukamoto H. Changes in methionine adenosyltransferase and S-adenosylmethionine homeostasis in alcoholic rat liver. *Am J Physiol*. 279:g178-g185 (2000).

Luger K, Mader AW, Richmond RK, Sergeant DF, Richmond TJ. Crystal structure of the nucleosome core particle at 2.8 A resolution. *Nature*, 398, 251-260 (1997).

Luo J. GSK3beta in ethanol neurotoxicity. Mol. Neurobiol. 40:108-121 (2009).

Lustberg MB, Ramaswamy B. Epigenetic targeting in breast cancer: therapeutic impact and future direction. *Drug News Perspect*. 22:369-381(2009).

MacDonald JL, Roskams AJ. Epigenetic regulation of nervous system development by DNA methylation and histone deacetylation. *Prog Neurobiol*. 88:170-183(2009).

Manning MA. Eugene HH. Fetal alcohol spectrum disorders: a practical clinical approach to diagnosis. *Neurosci. Biobehav. Rev*. 31:230-238 (2007).

Mason JB, Choi SW. Effects of alcohol on folate metabolism: implications for carcinogenesis. *Alcohol* 35:235-241(2005).

Marks PA, Xu WS. Histone deacetylases inhibitors: potential in cancer therapy. *J Cell Biochem*. 107:600-608(2009).

Martinez-Balbas MA, Bauer UM, Nielsen SJ, Brehm A, Kouzarides T. Regulation of E2F1 activity by acetylation. *EMBO J.*19: 662-671 (2000).

May PA, Gossage JP. Estimating the prevalence of fetal alcohol syndrome. A summary. Alcohol Res Health, 25:159-167(2001).

McGeer PL, McGeer EG, Nagai T. GABAergic and cholinergic indices in various regions of rat brain after intracerebral injections of folic acid. *Brain Res*. 260:107-116(1983).

McLaughlin F, La Thangue NB. Histone deacetylases inhibitors open new doors in cancer therapy. *Biochem. Pharmacol*. 68, 1139-1144 (2004)

Minard ME, Jain AK, Barton MC. Analysis of epigenetic alterations to chromatin during development. *Genesis*, 47: 559-572 (2009).

Mudbhary R, Sadler KC. Epigenetics, Development, and cancer: zebrafish make their mark. *Birth Defects Res*. 93C:194-203 (2011).

Marutha Ravindran CR, Ticku MK. Role of CpG islands in the upregulation of NMDA receptor NR2B gene in cortical neurons after chronic ethanol treatment in mice. *Mol Brain Res*. 121:19-27(2004).

Marutha Ravindran CR, Ticku MK. Role of CpG islands in the upregulation of NMDA receptor NR2B gene expression following chronic ethanol treatmentof cultured cortical neurons of mice. *Neurochem Int*. 46:313-327(2005).

Medina-Franco JL, Caulfield T. Advances in the computational development of DNA methyltransferase inhibitors. *Drug Discov. Today*, 16:418-425 (2011).

Meeran SM, Ahmed A, Tollefsbol TO. Epigenetic targets of bioactive dietary components for cancer prevention and therapy. *Clin Epigenetics*, 1: 101-116 (2010).

Mistry AR, Pedersen EW, Solomon E, Grimwade D. The molecular pathogenesis of acute promyelocytic leukaemia: implications for the clinical management of the disease. *Blood Rev*. 17: 71-97 (2003).

Momparler RL. Cancer epigenetics. *Oncogene*, 22:6479-6483(2003).

Mudbhary R, Sadler KC. Epigenetics, development, and cancer: zebrafish make their mark. *Birth Defects* Res. 93:194-203(2011).

Nash K, Rovet J, Greenbaum R, Fantus E, Nulman I, Koren G. Identifying the behavioral phenotype in fetal alcohol spectrum disorder: sensitivity, specificity and screening potential, *Arch Womens Ment Health*. 9:181-186 (2006).

O'Gara M, Klimasauskas S, Roberts RJ, Cheng X. Enzymatic C5-cytosine methylation of DNA: mechanistic implications of new crystal structures for Hhal methyltransferase-DNA-AdoHcy complexes. *J Mol Biol*. 261:634-645 (1991)

Okano M, Bell DW, Haber DA, Li E. DNA methyl transferases Dnmt3a and dnmt3b are essential for de novo methylation and mammalian development. *Cell*, 99:247-257 (1999).

Ouko LA, Shantikumar K, Knezovich J, Haycock P, Schnugh DJ, Ramsay M. Effect of alcohol consumption on CpG methylation in the differentially methylated regions of H19 and IG-DMR in male gamates: implications for fetal alcohol spectrum disorders. *Alcohol Clin Exp Res*. 33:1615-1627(2009).

Pandey SC, Ugale R, Zhang H, Tang L, Prakash A. Brain chromatin remodeling: a novel mechanism of alcoholism. *J Neurosci*. 28:3729-3737(2008).

Park PH, Miller R, Shukla SD. Acetylation of histone H3 at lysine 9 by ethanol in rat hepatocytes. *Biochem Biophys Res Commun*. 306:501-504(2003).

Park PH, Lim RW, Shukla SD. Gene-selective histone H3 acetylation in the absence of increase in global histone acetylation in liver of chronically fed alcohol. *Alcohol alcohol*. 2012 (in press).

Peterson CL, Laniel MA. Histones and histone modifications. *Curr. Biol*. 14: R546-R551 (2004).

Pfeifer GP, Tanguay RL, Steigerwald SD, Riggs AD. In vivo footprint and methylation analysis by PCR-aided genomic sequencing: comparison of active and inactive X chromosal DNA at the CpG island and promoter of human PGK-1. *Genes Dev*. 4:1277-1287(1990).

Pietrzykowski AZ, Friesen RM, Martin GE, Puig SI, Nowak CL, Wynne PM, Siegelmann HT,Treistman SN. Posttranscriptional regulation of BK channel splice varient stability by miR-9 underlies neuroadaptation to alcohol. *Neuron*, 52:274-287 (2008).

Portela A, Esteller M. Epigenetic modifications and human disease. *Nat Biotechnol.* 28:1057-1068(2010)

Rani CS, Qiang M, Ticku MK. Potential role of cAMP response element-binding protein in ethanol-induced N-methyl-D-aspartate receptor 2B subunit gene transcription in fetal mouse cortical cells. *Mol Pharmacol.* 67:2126-2136(2005).

Renthal W, Nestler EJ. Histone acetylation in drug addiction. *Semin Cell Dev Biol.* 20:387-394 (2009).

Rhee I, Bachman KE, Park BH, Jair KW, Yen RW, Schuebel KE, Cui H, Feinberg AP, Lengauer C, Kinzler K.W., Baylin SB, Vogelstein B. DNMT1 and DNMT3b cooperate to silence genes in human cancer cells, *Nature*, 416(6880) 552-556(2002).

Robertson KD, Ait-Si-Ali S, Yokochi T, Wade PA, Jones PL, Wolffe AP. DNMT1 forms a complex with Rb, E2F1 and HDAC1 and represses transcription from E2F-responsive promoters. *Nat. Genet.* 25:338-342 (2000).

Roth SY, Denu JM, Allis CD. Histone acetyltransferases. *Annu. Rev. Biochem.* 70: 81-120 (2001).

Rountree MR, Bachman KE, Baylin SB. DNMT1 binds HDAC2 and a new co-repressor, DMAP1, to form a complex at replication foci. *Nat Genet,* 25, 269-277 (2000).

Sathyan P, Golden HB, Miranda RC. Competing interactions between micro-RNAs determine neural progenitor survival and proliferation after ethanol exposure: evidence from an ex vivo model of the fetal cerebral cortical neuroepithelium. *J Neurosci* 27:8546-8557(2007).

Sato F, Tsuchiya S, Meltzer SJ, Shimizu K. MicroRNAs and epigenetics. *FEBS J.* 278:1598-1609(2011).

Schaefer M, Lyko F. Solving the Dnmt2 enigma. Chromosoma, 119:35-40(2010).

Schubeler D, MacAlpine DM, Scalzo D, Wirbelauer C, Kooperberg C, van Leeuwen F, Gottschling DE, O'Neill LP, Turner BM, Derlow J, Bell SP, Groudine M. The histone modification pattern of active genes revealed through genome-wide chromatin analysis of a higher eukaryote. *Genes Dev.* 18:1263-1271 (2004).

Setoguchi H, Namihira M, Kohyama J, Asano H, Sanosaka T, Nakashima K. Methyl-CpG binding proteins are involved in restricting differentiation plasticity in neurons. *J Neurosci Res.* 84:969-979(2006).

Shi Y, Lan F, Matson C, Mulligan P, Whetstine JR, Cole PA, Casero RA, Shi Y. Histone demethylation mediated by the nuclear amine oxidase homolog LSD1. *Cell*, 119: 941-953 (2004).

Shilatifard A. Chromatin modifications by methylation and ubiquitination: implications in the regulation of gene expression. *Annu Rev Biochem* 75: 243-269 (2006).

Shukla SD, Velazquez JV, French SW, Lu SC, Ticku MK, Zakhari S. Emerging role of epigenetics in the actions of alcohol. *Alcohol Clin Exp Res.* 32:1525-1534(2008).

Sims RJ 3rd, Reinberg D. Histone H3 Lys 4 methylation: caught in a bind? *Genes Dev* 20:2779-2786(2006).

Singal R, Ginder GD. DNA methylation. *Blood* 93:4059-4070(1999)

Soares AR, Pereira PM, Ferreira V, Reverendo M, Simoes J, Bezerra AR, Moura GR, Santos MAS. Ethanol exposure induces up-regulation of specific microRNAs in zebrafish embryos. *Toxicol. Sci.* (in press) (2012).

Sokol RJ, Delaney-Black V, Nordstrom VB. Fetal alcohol spectrum disorder. *JAMA* 290:2996-2999 (2003).

Spadoni AD, McGee CL, Fryer SL, Riley EP. Neuroimaging and fetal alcohol spectrum Disorders, *Neurosci. Biobehav. Rev*. 31:239-245 (2007).

Strahl BD, Allis CD. The language of covalent histone modifications. *Nature*, 403: 41-45 (2000).

Streissguth AP, O'Malley K. Neuropsychiatric implications and long-term consequences of fetal alcohol spectrum disorders. *Semin. Clin. Neuropsychiatry*. 5:177-190 (2000).

Suzuki T, Miyata N. Epigenetic control using natural products and synthetic molecules. *Curr Med Chem*. 13: 935-958 (2006).

Sullivan, W.C., 1899. A note on the influence of maternal inebriety on the offspring. *J. Ment. Sci*.45, 489-503.

Takizawa T, Nakashima K, Namihira M, Ochiai W., Uemura A, Yanagisawa M, Fujita N, Nakao M, Taga T. DNA methylation is a critical cell-intrinsic determinant of astrocyte differentiation in the fetal brain. *Dev Cell*. 1:749-758(2001).

Tahiliani M, Koh KP, Shen Y, Pastor WA, Bandukwala H, Brudno Y, Agarwal Suneet, Iyer LM, Liu DR, Aravind L, Rao A. Conversion of 5-methylcytosine to 5-hydroxymethylcytosine in mammalian DNA by MLL partner TET1. *Science*, 324(5929): 930-535 (2009).

Thanh NX, Jonsson E. Costs of fetal alcohol spectrum disorder in Alberta, Canada. *Can J Clin Pharmacol*. 16:e80-90(2009).

Tilghman SM. The sins of the fathers and mothers: genomic impriting in mammalian development. *Cell*, 96:185-193(1999).

Tsukada Y, Fang J, Erdjument-Bromage H, Warren ME, Borchers CH, Tempst P, Zhang Y. Histone demethylation by a family of JmjC domain-containing proteins. *Nature*, 439: 811-816 (2006).

Travers AA, Thompson JM. An introduction to the mechanics of DNA. Philos Transact. *A Math. Phys. Eng. Sci*. 362:1265-1279 (2004).

Ueda Y, Okano M, Williams C, Chen T, Georgopoulou K, Li E. Roles for Dnmt3b in mammalian development: a mouse model for the ICF syndrome. *Development*, 133:1183-1192 (2006).

Urban M, Chersich MF, Fourie LA, Chetty C, Olivier L, Viljoen D. Fetal alcohol syndrome among grade 1 schoolchildren in Northern Cape Province: Prevalance and risk factors. *S Afr Med J*. 98:877-882(2008).

Vaissiere T, Sawan C, Herceg Z. Epigenetic interplay between histone modifications and DNA methylation in gene silencing. *Mutat. Res*. 659: 40-48 (2008).

Veeck J, Esteller M. Breast cancer epigenetics: from DNA methylation to microRNAs. *J. Mammary Gland Biol. Neoplasia*, 15, 5-17 (2010).

Viljoen DL, Gossage JP, Brooke L, Adnams CM, Jones KL, Hoyme HE, Snell C, Khaolc NC, Kodituwakku P, Asante KO, Findlay R, Quinton B, Marais AS, Kalberg WO, May PA. Fetal alcohol syndrome epidemiology in a South African community: a second study of a very high prevalence area. *J Stud Alcohol*, 66:593-604(2005).

Wang G, Wang X, Wang Y, Yang JY, Li L, Nephew KP, Edenberg HJ, Zhou FC Liu Y. Identification of transcription factor and miceroRNA binding sites in responsible to fetal alcohol syndrome. *BMC Genomics*, 9:S19 (2008).

Wang L-L, Zhang Z, Li Q, Yang R, Pei X, Xu Y, Wang J, Zhou S-F, Li Y. Ethanol exposure induces differential MicroRNA and tagget gene expression and teratogenic effects which can be suppressed by folic acid supplementation. *Human Reprod*. 24:562-579(2009).

Weber B, Stresemann C, Brueckner B, Lyko F. Methylation of human MicroRNA genes in normal and neoplastic cells. *Cell Cycle,* 6:1001-1005 (2007).

Wilson MJ, Shivapurkar N, Poirier LA. Hypomethylation of hepatic nuclear DNA in rats fed with a carcinogenic methyl-deficient diet. *Biochem J.* 218:987-990(1984).

Wolffe AP, Matzke MA. Epigenetics: regulation through repression. *Science*, 286 (5439)481-486(1999)

Wu H, Coskun V, Tao J. Xie W, Ge W, Yoshikawa K, Li E, Zhang Y, Sun YE. Dnmt3A-dependent nonpromoter DNA methylation facilitates transcription of neurogenic genes. *Science*, 329:444-448 (2010).

Wilson MJ, Shivapurkar N, Poirier LA. Hypomethylation of hepatic nuclear DNA in rats fed with a carcinogenic methyl-deficient diet. *Biochem J.* 218:987-990(1984)

Wu JC, Santi DV. Kinetic and catalytic mechanism of Hhal methyltransferase. *J. Biol. Chem.* 262, 4778-4786 (1987).

Xu P, Guo M, Hay BA. MicroRNAs and the regulation of cell death. *Trends Genet.* 20: 617-624(2004).

Yamane K, Tateishi K, Klose RJ, Fang J, Fabrizio LA, Erdjument-Bromage H, Taylor-Papadimitriou J, Tempst P, Zhang Y. PLU-1 is an H3K4 demethylase involved in transcriptional repression and breast cancer cell proliferation. *Mol Cell*, 25:801-812 (2007).

Yoo CB, Jones PA. Epigenetic therapy of cancer: past, present and future. *Nat Rev Drug Discov*. 5:37-50(2006).

Zeisel SH. What choline metabolism can tell us about the underlying mechanisms of fetal alcohol spectrum disorders. *Mol Neurobiol*. 44:185-191(2011).

Zhou FC, Balaraman Y, Teng M, Liu Y, Singh RP, Nephew KP. Alcohol alters DNA methylation patterns and inhibits neural stem cell differentiation. *Alcohol Clin Exp Res.* 35:735-746(2011).

In: Ethanol: Production, Cellular Mechanisms and Health Impact ISBN: 978-1-62257-297-7
Editors: Roy I. Henry and Billy P. Woods

Chapter 5

AN ANALYSIS OF ETHANOL INVESTMENT DECISIONS

C.-Y. Cynthia Lin
University of California at Davis, Davis, California, US

ABSTRACT

Recently the support of biofuel production has been a politically sensitive topic. Several government policies actively promote ethanol production via tax incentives and mandates. Additionally, politicians have pushed for ethanol as an environmentally friendly alternative to imported oil, as well as a way to boost farm profits and improve rural livelihoods. All these factors have coincided with investment in the industry in recent years. Even when excluding the U.S., which was the country with the largest fuel-ethanol production in 2009, the fuel-ethanol industry has been growing rapidly in the rest of the world (ROW). Ethanol-producing countries in the ROW include Brazil, Canada, China, and Thailand, as well as countries in Europe. There are 191 fuel-ethanol plants in the ROW, which is a little more than in the U.S., and 82% of them were built after 2005.

This chapter reviews some of the papers I have written analyzing what factors affect the decision to invest in building new ethanol plants using a dynamic structural econometric model of the investment timing game. The results of my research will help determine which policies and factors can promote fuel-ethanol industry development.

In Lin and Thome (2012), we estimate a model of the investment timing game in corn ethanol plants in the United States. This model follows my previous work estimating a structural econometric model of the multi-stage dynamic investment timing game in offshore petroleum production (Lin, 2012). In Lin and Yi (2012a,b), we estimate a model of the investment timing game in ethanol plants worldwide that allows for the choice among different feedstocks.

According to our results, we find that, in the United States, competition between plants is enough to deter local investments. We also find that availability of feedstock is important in determining plant location. We also find that in the United States state producer tax credits and the federal MTBE ban have a positive effect on ethanol investment (Lin and Thome, 2012). In Europe, competition between plants deters local investments and ethanol support policies encourage investments (Lin and Yi, 2012b). In Canada, competition between plants is enough to deter local investments, the availability of feedstock is important in determining plant location, and the effects of policy support for wheat-based plants are significant (Lin and Yi, 2012a).

INTRODUCTION

Recently the support of biofuel production has been a politically sensitive topic. Politicians have pushed for support for fuel ethanol production as an environmentally friendly alternative to imported oil, as well as a way to boost farm profits and improve rural livelihoods. Several government policies actively promote ethanol production via tax incentives and mandates, and these policies are blamed for rising food prices around the world (Mitchell, 2008). It is important to understand the factors that have motivated the significant local investments in the ethanol industry that have been made since the mid-1990s both in the U.S. and worldwide.

Fuel ethanol has been in use in the United States since the time of the Model T Ford (the original flex-fuel vehicle), and while the United States passed Brazil in ethanol production in 2005, today ethanol is mostly relegated to status as a gasoline additive. The first US ethanol boom began as a result of the oil embargoes in 1973 and 1979. The desire for more energy self-sufficiency, the resulting legislation (in the form of federal income tax credits and blender's credits that continue today), and the phase out of leaded gasoline led to the construction of 153 new plants by 1985 (DOE, 2008). These plants were tiny by today's standards, with an average capacity of 8 million gallons per year, and by 1991 only 35 were still operational due to poor business judgment and bad engineering (DOE, 2008; Urbanchuck, 2006).

The second US ethanol boom began in the mid-1990s and hit full-stride by the early 2000s. Several factors contributed to this most recent boom. The Clean Air Act of 1990 mandated use of oxygenates in gasoline, of which ethanol is one, and the subsequent phase out and ban of MTBE as additive beginning in the late 1990s further increased demand for ethanol. Additionally the Renewable Fuel Standards of the Energy Policy Act of 2005 mandated ethanol production floors beginning in 2007, which rise to 36 billion gallons per year in 2033. Over this time period, the number of ethanol plants rose from 35 plants in 1991, to 50 in 1999, to 192 in September of 2010 for a total capacity of 13 billion gallons per year.

In addition to the policy and demand-side contributors to the recent ethanol boom, this new industry growth has been accompanied by changes in plant management and technology. Most significantly, the average capacity of plants in our focus region was 62 million gallons per year in 2008 up from 8 million gallons per year in 1985. In the mid-1990s the industry began designing more efficient plants, which use natural gas instead of coal as fuel (DOE, 2008). Ownership is also shifting to streamlined corporate owners with multiple plants. Historically, farmer-owned plants had a large share of the market, though by 2007 only 11% of new capacity was farmer owned, while the largest 5 corporations had 42% of capacity in 2008 (FTC, 2008).

This recent boom, in addition to industry changes in technology and ownership structure, begs an analysis of investment decisions. Most ethanol plants use corn as a feedstock, and thus are located in the Midwestern United States, where the majority of the corn in the US is grown; these plants are the focus of our study. Since biofuels have been touted as a way to enhance profits in rural areas, where grain prices have remained stagnant over time, it is important to determine what factors affect decisions about when and where to invest in building new ethanol plants. In Lin and Thome (2012), we model this decision in this paper using both reduced-form and structural models.

Even when excluding the U.S., which was the country with the largest fuel-ethanol production in 2009, the fuel-ethanol industry has been growing rapidly in the rest of the world (ROW). Ethanol-producing countries in the ROW include Brazil, Canada, China, and Thailand, as well as countries in Europe. There are 191 fuel-ethanol plants in the ROW, which is a little more than in the U.S., and 82% of them were built after 2005. In Lin and Yi (2012a,b), we estimate a model of the investment timing game in ethanol plants worldwide that allows for the choice among different feedstocks.

In Europe, 20 countries have fuel ethanol production and most of the fuel ethanol plants were built after 2000. The development of European biofuel is based on two Directives: the Renewable Energy Directive (RED) of 2003/30/EC sets indicative targets of 2% renewable fuels in transport by 2005 and 5.75% by 2010 but is not legally binding, and the RED of 2009/28/EC is made mandatory and therefore legally binding. The main policies fuel ethanol policies in Europe include a tax credit, a blending mandate and R&D support. Most of the policies were implemented after 2003. Empirical research shows that the effects of policies for the U.S. fuel ethanol production are positive (Lambert et al., 2008; Sarmiento and Wilson, 2007; Lin and Thome, 2012), however, whether the stimulation effects of the government policies play the same role in Europe is not yet clear, especially for the different varieties of feedstocks.

There is a related literature on food manufacturing location decisions which begins with a basic model of determinants of manufacturing establishment growth. One example is Goetz's (1997) analysis of the determinants of rural food manufacturing establishment growth. He considers the effects of the following factors: access to output markets, labor force composition and quality, transportation infrastructure, government intervention, and availability of raw materials. In his model, location decisions involve a two-step process where regions are first chosen for broader consideration, and then the choice is narrowed within each region.

The decision to build a fuel ethanol plant involves a dynamic decision-making process. Investors have to consider the entire stream of per-period payoffs from now into the future. Because values of the state variables can change over time, for example if ethanol and feedstock prices change, or if there is a chance that neighboring plants might be constructed and start production, there is an option value to waiting that makes the decision dynamic rather than static.

When the decision of an investor is affected by neighbors' investment decisions, it becomes a more complicated decision-making problem. There are two types of effects that add a strategic (or non-cooperative) dimension to the potential entrants' investment timing decisions. The first type of effect is a *competition effect*, which arises if there is more than one ethanol plant located in one region so that these plants compete in feedstock supply when they choose the same feedstock or compete in local fuel ethanol market given limited demand[1]. The competition effect deters ethanol plants for entering in regions where there are other ethanol plants already present. The second type of effect is an *agglomeration effect*: if

[1] To produce fuel ethanol, the transportation costs of bulky grains or sugarbeets take a significant share of the feedstock costs, and around 70% of the cost is from feedstock for ethanol production (Whittington, 2006). Therefore, the distance from a plant to the feedstock production area is extremely important. Usually, fuel ethanol transportation is harder and more expensive than gasoline transportation due to the fact that ethanol can easily absorb water during the transportation process, and thus is normally transported using tank trucks for dehydrated fuel ethanol. The higher transportation cost prevents the ethanol from being sold too far away.

there are several ethanol plants located in the same region, the existing plants may have developed transportation and marketing infrastructure and/or an educated work force that new plants can benefit from. The agglomeration effect induces an ethanol plant to locate near other plants, since a fuel ethanol plant benefits from the existence of other plants. Owing to both competition and agglomeration effects, the dynamic decision-making problem faced by the potential ethanol plants is not merely a single-agent problem, but rather can be viewed as a non-cooperative game in which plants behave strategically and base decisions on their neighbors' strategies.

There are few studies that specifically address the location determinants of ethanol plants in the United States and this is the first that econometrically models the entry location decision.[2] Sarmiento and Wilson (2007) use a cross-sectional discrete choice model to analyze agricultural characteristics and spatial dimensions that determine plant location. Similarly, Lambert et al. (2008) use a cross-sectional discrete choice model spatial clustering to look at factors that affect the presence of ethanol plants and proposed plants in a given county, and also isolate clusters that may attract investment. Finally, Haddad et al. (2010) model state-by-state spatial determinants of plant location, and find significant differences across states.

While these studies analyze similar factors as Goetz (1997), they are cross-sectional and so cannot address investment timing decisions and thus only focus on location determinants, providing a starting point for our analysis as far as identifying potentially important exogenous factors. However, the results of these studies are not always qualitatively similar because of the different empirical specifications and regional foci. For example, the first two studies find that access to corn is an important location determinant. However, Haddad et al. (2010) do not find access to corn significant, though they note that following location theory (see Goetz, 1997), firms might choose a region with a lot of corn, and in the second stage will make their location decision based on other factors; this two-stage decision-making process is why they find differences in location determinants across states.

None of the studies adequately addresses the potential competition between plants in the location decision. Haddad et al. (2010) do not include potential spatial competition between plants as an explanatory variable, perhaps since most competing plants are not located in the same county. Lambert et al. (2008) include plants established before 2000 as an explanatory variable and find a negative impact on new plants and announcements between 2000 and 2007, though there is no analysis of spatial relationship or relative timing on those (potential) entrants. Sarmiento and Wilson (2007) employ a cross-sectional model of plant location with a spatially lagged dependent variable in order to estimate the competitive effect between plants. They find a large negative effect of a nearby plant on the probability of another plant locating nearby, and furthermore, that this effect decreases with distance. It is important to note however that these competitive effects only describe the relationship between existing plants; neither of these models has a time element and without panel data it is not possible to analyze the effect of competition on entry.

This chapter reviews some of the papers I have written analyzing what factors affect the decision to invest in building new ethanol plants using a dynamic structural econometric

---

[2] The real options literature models entry and exit thresholds, but cannot explicitly address location (see e.g., Schmidt et al., 2009). Most of the literature about the US ethanol industry focuses on impacts on other markets. For example, McNew and Griffith (2005) look at impacts on local corn markets.

model of the investment timing game. This work improves upon the previous literature by estimating a dynamic model with panel data, and by directly estimating the effect of covariates on the investment decision itself. The results of my research will help determine which policies and factors can promote fuel-ethanol industry development.

METHODOLOGY

In Lin and Thome (2012), we first estimate a reduced-form discrete choice model by regressing the probability of investment in an ethanol plant on the covariates using a fixed-effects logit model. It is logical to begin with the fixed-effects model since unobservable county characteristics might explain their ability to attract investment in an ethanol plant. The major advantage of a fixed-effects model is the addition of the county-specific effect. This allows us to control for unobservable county traits, such as openness to or promotion of business, that remain fixed over time. This is particularly important since the resolution of our data is not ideal and some variables are not observed at the county level. We next specify the strategic variable, other plants, as capacity instead of count. We hypothesize that larger or smaller competing capacities could have different effects. Third, we group the competing plants by type: cooperatively-owned, ethanol-only firms, and conglomerates. We hypothesize that different types of operators may produce different externalities (either positive and negative) towards potential entrants. We estimate a fourth model using entry of different type of plant (coop, ethanol-only, and conglomerate) as the dependent variable in a pooled multinomial logit model.

In Lin and Thome (2012), we follow up a reduced-form model of investment in an ethanol plant with a structural model. We use a structural model for several reasons. First, it is interesting to estimate the effect of the state variable the expected payoff from investing in an ethanol plant. In the reduced-form model, we estimate the effect of these variables on the probability of investment. Second, the structural model makes it possible to estimate the strategic interaction between different regions that produce corn and could invest in ethanol plants. Thirdly, the structural model explicitly models the dynamic investment decision, including the continuation value to waiting. In contrast, the reduced-form model only estimates the per-period probability of investment.

As explained by Reiss and Wolak (2007), a structural econometric model is one that combines economic theory with a statistical model, enabling us to estimate structural parameters. Incorporating firm dynamics into structural econometric models enhances our understanding of behavior and also enables us to estimate structural parameters which have a transparent interpretation within the theoretical model that frames the empirical investigation (Aguirregabiria and Mira, 2010).

Dynamic discrete choice structural models are useful tools in the analysis of economic and social phenomena whenever strategic interactions are an important aspect of individual behavior. In the ethanol market, because a firm's costs and market demand hinge on the structure of market, a firm's decision depends on its conjecture about competitors' behavior. This type of model assumes agents are forward looking and maximize the expected discounted value of the entire stream of payoffs. Agents are assumed to make decisions based only on historic information directly related to current payoffs, and history only influences current decisions insofar as it impacts a state variable that summarize the direct influence of the past on current payoffs.

Recent papers have developed techniques for estimating dynamic games between multiple agents. Pakes, Ostrovsky, and Berry (2007) illustrated a dynamic entry/exit game where the structural parameters could be estimated semi-parametrically. Bajari, Benkard, and Levin (2007) added to this literature by describing the estimation of the parameters in a dynamic game with continuous control variables.

In Lin and Thome (2012), we estimate a model of the investment timing game in corn ethanol plants in the United States. This model follows my previous work estimating a structural econometric model of the multi-stage dynamic investment timing game in offshore petroleum production (Lin, 2012), which is based on an econometric model developed by Pakes, Ostrovsky and Berry (2007). In Lin (2012), I build on the work of Pakes, Ostrovsky and Berry (2007) on discrete games of entry and exit by examining sequential investments with a finite horizon. The econometric estimation technique takes place in two steps. In the first step, the continuation value is estimated nonparametrically and used to form the model's estimate of the investment probabilities. In the second step, the investment probabilities predicted by the model are matched with the empirical investment probabilities in the data using generalized method of moments.

In Lin and Yi (2012a,b), we estimate a model of the investment timing game in ethanol plants worldwide that allows for the choice among different feedstocks. This research differs from previous studies of the investment and location of ethanol plants because it models the decision as a dynamic one rather than a static one, because it allows for the choice among multiple feedstocks rather than just one feedstock such as corn, because its strategic framework allows the estimation of strategic interactions among plants, and because it uses international data rather than data from the U.S.

CONCLUSION

The results of both the reduced-form and structural estimation in Lin and Thome (2012) indicate that there is an important strategic component to investment in ethanol plants. This net-negative effect may be due to localized competition, though we also find that plant identity matters as far as the strength and sign of this effect. We cannot explore the plant identity question (or exit) more in our current framework, but this is subject of ongoing work.

In Lin and Thome (2012), we also find that intensity of corn production is important in determining local investment in both models. Corn is bulky and transportation is not cheap, so it is beneficial for plants to locate where they have good access to feedstock.

In Lin and Thome (2012), we find mixed results of the effects of input and output prices across the different models and specifications. This inconsistency is potentially due to the data resolution, which, at state and national level, is not ideal. That said, we were still able to find some effect of prices indicating that 1) they do matter, and 2) even with less than ideal data source, our model is strong enough to tease out some of these effects.

In Lin and Thome (2012), we also find mixed results of the policy variables. In the reduced-form specification, we find that the state producer tax breaks are important, and the MTBE ban is significant in the structural model. The differences here may be due to the models themselves. The MTBE ban affects the market via increased ethanol demand and higher expected prices; thus in the structural model it leads to increased expected profits from investment. The producer support/tax break only applies to some plants, and only to some of the production, so we may not have identified this effect in the structural model, where the profit function is the same for each county. Further analysis of the effects of policies on investment in ethanol plants is subject of ongoing work.

In Lin and Yi (2012a,b), we first construct a dynamic discrete choice model for a potential fuel ethanol plant in which the investor maximizes its present discounted value of its entire stream of payoffs, and in which the decisions of other plants in the same local market affect an investor's decision.

The innovative features of our model are the consideration of interactions between fuel ethanol plants and the dynamic decision making framework. Once the dynamic decision making process has been fully described, the effects of economic, policy and strategic variables on profit can be estimated through a semiparametric approach.

Our results in Lin and Yi (2012b) show that the potential investor considers various exogenous conditions: higher ethanol prices, gasoline prices, plenty of feedstocks and government support are very helpful to improve the profits, while a high natural gas price would harm the profits although some of the coefficients on natural gas are not significant. The negative competition effects of livestock vary across the choices of feedstocks. There is a negative effect on profits for all types of fuel ethanol plants which come from existing plants. In addition, the strategic interactions resulting from competition and agglomeration effects from other potential entrants are not obvious.

Thus, according to our results, we find that, in the United States, competition between plants is enough to deter local investments. We also find that availability of feedstock is important in determining plant location. We also find that in the United States state producer tax credits and the federal MTBE ban have a positive effect on ethanol investment (Lin and Thome, 2012). In Europe, competition between plants deters local investments and ethanol support policies encourage investments (Lin and Yi, 2012b). In Canada, competition between plants is enough to deter local investments, the availability of feedstock is important in determining plant location, and the effects of policy support for wheat-based plants are significant (Lin and Yi, 2012a).

References

Aguirregabiria, V., and Mira, P. (2010). "Dynamic discrete choice structural models: A survey," *Journal of Econometrics*, 156(1): 38-67.

Bajari, P., C. L. Benkard, J. Levin (2007). "Estimating Dynamic Models of Imperfect Competition". *Econometrica*, 75(5): 1331-70.

DOE [Deparment of Energy] (2008). Energy Time Lines: Ethanol. Revised June, 2008. Accessed January 2009.

FTC [Federal Trade Commission] (2008). *Report on Ethanol Market Concentration.*

Goetz, S. (1997). "State- and county-level determinants of food manufacturing establishment growth: 1987-93". *American Journal of Agricultural Economics*, 79, 838-850.

Haddad, M.A., G.Taylor, F. Owusu (2010). Locational choices in of the ethanol industry in the Midwest Corn Best. *Economic Development Quarterly*, 24(1): 74-86.

Lambert, D.M., M. Wilcox, A. English, and L. Stewart (2008). "Ethanol plant location determinants and county comparative advantage". *Journal of Agricultural and Applied Economics*, 40, 117-135.

Lin, C.-Y.C. (2012). Strategic decision-making with information and extraction externalities: A structural model of the multi-stage investment timing game in offshore petroleum production. *Working paper.* University of California at Davis.

Lin, C.-Y.C., and K. Thome. (2012). Investment in corn-ethanol plants in the Midwestern United States. *Working paper.* University of California at Davis.

Lin, C.-Y.C., and F. Yi, (2012a). Ethanol plant investment in Canada: A structural model. *Working paper.* University of California at Davis.

Lin, C.-Y.C. and F. Yi. (2012b). What factors affect the decision to invest in a fuel ethanol plant?: A structural model of the investment timing game in Europe. *Working paper.* University of California at Davis.

Mitchell, Donald (2008) A note on rising food prices. *Policy resrach Working Paper # 4862,* The World Bank Development Prospects Gorup, July 2008.

Pakes, A., M. Ostrovsky, S. Berry (2007). "Simple estimators for the parameters of discrete dynamic games (with entry and exit examples)". *RAND Journal of Economics, 38*(2): 373.

Reiss, Peter C. and Wolak, Frank A. (2007), "Structural Econometric Modeling: Rationales and Examples from Industrial Organization," *Handbook of Econometrics*, Volume 6A, (edited by James J. Heckman and Edward E. Leamer), *6*:4277-4415.

Sarmiento, C., and W.W. Wilson. (2008). *"Spatial competition and ethanol plant location decisions,"* 2008 Annual Meeting, July 27-29, 2008, Orlando, Florida 6175, American Agricultural Economics Association.

Urbanchuk, J. M. (2006). *"Economic Impacts on the Farm Community of Cooperative Ownership of Ethanol Production".* National Corn Growers Association Report.

In: Ethanol: Production, Cellular Mechanisms and Health Impact ISBN: 978-1-62257-297-7
Editors: Roy I. Henry and Billy P. Woods

Chapter 6

Cellulosic Ethanol Production through Bioconversion of Lignocellulosic Biomass

Yi Zheng *
Biological and Agricultural Engineering Department,
University of California, California, US

Abstract

Cellulosic ethanol has been taken as a promising alternative energy source for the nonrenewable fossil fuel. Cellulose is one of the three primary polymers of lignocellulosic biomass cell wall. It is the most abundant biomaterial on earth. Each cellulose molecule is a linear polymer of thousands of glucose residues. Cellulose can be hydrolytically broken down into glucose either enzymatically by cellulolytic enzymes or chemically by sulfuric or other acids. The produced glucose can be fermented into ethanol. However, structural and compositional features of lignocellulosic biomass such as lignin/hemicellulose shield and crystallinity of cellulose present challenges to efficient hydrolysis of cellulose. Employment of enzymes for the hydrolysis of the lignocellulose is considered the prospectively most viable strategy to offer advantages over other chemical conversion routes of higher yields, minimal byproduct formation, low energy requirements, mild operating conditions, and environmentally friendly processing. It has long-term potential for cost reductions compared to other more established routes such as concentrated acid and two-stage dilute acid hydrolysis, even though the enzymatic route has the highest current costs. Considerable research efforts have been made to improve the bioconversion efficiency of cellulose into ethanol, mainly focusing on three major steps of biomass pretreatment, enzymatic hydrolysis, and fermentation. This chaper reviews the development of biochemical conversion of biomass materials into cellulosic ethanol, including previous achievements, current status, and future work.

* Corresponding author: Tel.: +1 530 752 8039; Fax: +1 530 752 2640. E-mail address: yizheng@ucdavis.edu (Y. Zheng).

INTRODUCTION

With the dramatic growth of world population and the progress of industrialization in more and more countries, the energy consumption has increased steadily over the last century. Crude oil has been the major resource to meet the increased energy demand. Crude oil is nonrenewable and its distribution highly depends on the geological conditions. The available crude oil reserve is depleting yearly. It was predicted that annual global oil production would decline to approximately 5 billion barrels in 2050 (Campbell and Laherrere, 1998). The consequences of inadequate oil availability could be severe because the economy in the US and much of the world depends on oil. The largest portion of the energy used in this country is derived from petroleum, approximately 60% of which is imported (Energy Information Administration, EIA, 2009). Our high dependence on imported oil has resulted in severe price shocks and shortages which has caused "Energy Crises" (actually "Petroleum Crises") during the 1970s because a number of oil-rich countries, especially in the Middle East, teamed together to form the oil cartel OPEC and manipulate the quantity and therefore price of petroleum (Wyman, 1999). Additionally, the utilization of fossil fuel has caused various serious environmental problems, e.g., global climate change due to the green house gas emission related to the consumption of fossil fuel. Currently, the US and much of the world are faced with complex economic and environmental issues associated with energy use that must be addressed if we are to maintain and improve our lifestyles (Wyman, 1996). Therefore, there is a great interest in exploring alternative energy sources.

Ethanol that is produced from lignocellulosic biomass resources has the potential to become a cost-effective and environmentally benign substitute fuel for petroleum at a low price. Unlike fossil fuels, ethanol is a renewable energy source produced through fermentation of biomass-derived sugars. Ethanol is widely used as a transportation fuel and fuel additive in the US. The 3.4 billion gallons of ethanol produced in 2004 represent about 2% of gasoline consumption (Genomics, 2009). Using ethanol-blended fuel for automobiles can significantly reduce petroleum use and exhaust greenhouse gas emission (Wang et al., 1999). Ethanol is also a safer alternative fuel additive to methyl tertiary butyl ether (MTBE), which is a toxic chemical compound and has been found to contaminate groundwater. The US Environmental Protection Agency announced regulations to eliminate MTBE in gasoline (Browner, 2000). However, current commercially available fuel ethanol is produced from starch and/or sugar-based crops such as corn and sugarcane not from lignocellulosic materials. A rapid increase in such kind of ethanol may not be practical because corn or sugarcane production for ethanol will compete for the limited agricultural land needed for food and feed production. As an energy source, therefore, the ethanol cost is relatively high compared to fossil fuels. Alternatively, low-cost ethanol can be produced from domestically abundant and inexpensive sources of lignocellulosic biomass including agricultural and forestry residues, a variety of industrial waste streams, municipal solid wastes (MSW), woody and herbaceous dedicated energy crops grown on underutilized land, and solid animal waste. Unlike starch or sugars from starch or sugar-based crops, lignocellulosic materials primarily consist of cellulose, hemicellulose, and lignin. Depending on plant species and cell type, the dry weight of a cell wall typically consists of about 35-50% cellulose, 20-35% hemicellulose, and 10-25% lignin (Saha, 2004). Table 1 shows the contents of cellulose, hemicellulose, and lignin in common lignocellulosic biomass. Such biomass-based ethanol is conveniently named bioethanol in this chapter.

Table 1. Percent dry weight composition of various lignocellulosic materials

Lignocellulosic material	Cellulose (%)	Hemicellulose (%)	Lignin (%)	Reference
Rice straw	41	22	10	Lee (1997)
Wheat straw	40	27	23	Kristensen *et al. (2008)*
Corn cob	45	35	15	Sun and Cheng (2002)
Corn stover	39	20	15	Lee (1997)
Bagasse fiber	38	27	18	Lee (1997)
Rice hull	36	20	19	Lee (1997)
Sugar beet pulp	20	23	1	My data
Tomato pomace	8	4	10	My data
Switch grass	31	20	18	Wiselogel et al. (1996)
Jose Tall Wheatgrass	31	20	20	Zheng et al. (2007)
Creeping Wild Ryegrass	34	21	24	Zheng et al. (2007)
Coastal Bermuda grass	25	36	6	Reshamwala et al.(1995)
Popular	50	17	18	Wiselogel et al. (1996)
Pine wood	46	9	29	Wiselogel et al. (1996)
Eucalyptus	44	13	35	Zheng et al. (2007)
Douglas fir	50	18	28	Lee (1997)
Waste papers from chemical pulps	60–70	10–20	5–10	Sun and Cheng (2002)
Office paper	69	12	11	Wiselogel et al. (1996)
Newspaper	64	22	21	Lee (1997)
Municipal solid waste	44	5	24	Zheng et al. (2007)
Solid cattle manure	2–5	1–3	3–6	Dewes and Hünsche (1998)

The overall benefits of developing bioethanol technology have been previously stated by Wyman (2003): increased national energy security, reduction in green house gas emissions, conversion of solid wastes into valuable fuels, foundation of biorefinery, and macroeconomic benefits for rural communities and society at large. Although extensive research has been completed on conversion of lignocellulosic materials to bioethanol in the last two decades and the benefits have been realized on bench and/or pilot scale, commercialization and widespread application of bioethanol technology has not yet occurred. Previous research has demonstrated that lignocellulosic biomass requires extensive processes in order to realize the conversion of biomass into bioethanol. Theoretically speaking, biomass-to-bioethanol includes two processes: hydrolysis of cellulose in the lignocellulosic materials to fermentable sugars, and fermentation of the sugars to bioethanol.

The hydrolysis is usually catalyzed by acids or cellulase enzymes, and the fermentation is carried out by yeasts or bacteria (Sun and Cheng, 2002). However, the compositional and structural features of the lignocellulosic biomass cell wall that have been identified to affect the hydrolysis of cellulose. Such features include cellulose crystallinity, specific surface area of cellulose, degree of polymerization, cellulose protection by lignin, hemicellulose sheathing, and degree of hemicellulose acetylation (Chang and Holtzapple, 2000; Hsu, 1996; Hsu et al., 1980; Kim and Holtzapple, 2005, 2006; McMillan, 1994; Rydholm, 1965; Wenzel, 1970; Yang and

Wyman, 2004). The presence of these factors makes cellulose recalcitrant to hydrolytic enzymes, thus reducing the efficiency of the hydrolysis.

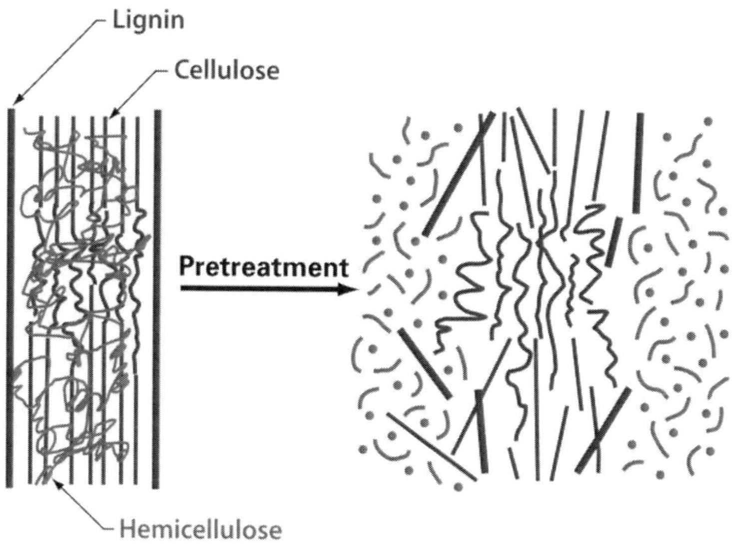

Figure 1. Schematic of goals of pretreatment on lignocellulosic material (adapted from Mosier et al., 2005).

In principal, an effective pretreatment or a combination of multiple pretreatments is expected to disrupt multiple barriers (Figure 1) so that hydrolytic enzymes can penetrate and perform hydrolysis on pretreated biomass more easily (McMillan, 1994).

OVERVIEW OF BIOETHANOL PRODUCTION

As mentioned above, pretreatment is required to alter the biomass macroscopic and microscopic size and structure as well as its submicroscopic structural and chemical composition to facilitate rapid and efficient hydrolysis of carbohydrates to fermentable sugars (Chang and Holtzapple, 2000). Although a variety of hydrolysis process configurations have been studied to realize the conversion of biomass into fermentable sugars, enzymatic hydrolysis of cellulose provides opportunities to improve the technology so that bioethanol is competitive to other liquid fuels on a large scale (Wyman, 1999). The resulted fermentable sugars from hydrolysis could be fermented into bioethanol by ethanologenic microorganisms, which can be either naturally occurring or genetically modified. Cellulose is usually organized into microfibrils, each measuring about 3 to 6 nm in diameter and containing up to 36 glucan chains having thousands of glucose residues. Depending on the degree of crystallinity, cellulose has two different statues in plant cell wall such as crystalline and paracrystalline (amorphous) cellulose. Hemicellulose, a branched polymer composed of pentose (5-carbon) and hexose (6-carbon) sugars, can be hydrolyzed by hemicellulases or acids into its component sugars, including xylose, arabinose, galactose, glucose and/or mannose, etc. Hexoses such as glucose, galactose, and mannose are readily fermented to ethanol by many naturally occurring organisms, but the pentoses including xylose and arabinose are fermented to ethanol by few native strains, and usually at relatively low yields.

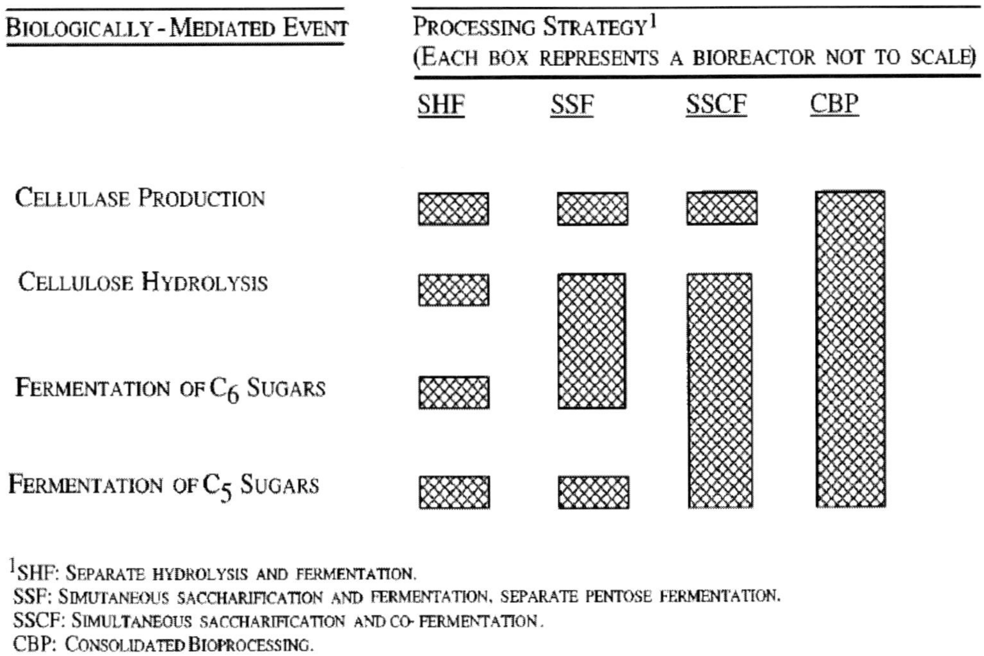

Figure 2. Consolidation of biologically mediated events in cellulosic ethanol production (adapted from Lynd, 1996).

Since xylose and arabinose in hemicellulose generally comprises a significant fraction of lignocellulosic biomass, especially hardwoods, agricultural residues and grasses (Wiselogel et al., 1996), it must be utilized to make economics of bioethanol processing feasible (Lynd, 1996). The development of genetically engineered microorganisms resulted in bacteria and yeasts capable of co-fermenting pentoses and hexoses into ethanol and other value-added products at high yields (Aristidou and Penttila, 2000; Bothast et al., 2002; Deanda et al., 1996; Dien et al., 1998, 2003; Ho et al., 1998, 1999; Ingram et al., 1998, 1999; Ingram and Doran, 1995; Jeffries and Shi, 1999; Mohagheghi et al., 1998; Ruohonen et al., 2006; Sues et al., 2005; Zhang et al., 1995).

Taking into the consideration of cellulolytic enzyme (i.e., cellulase) production and the process configuration of hydrolysis and fermentation, four biologically mediated steps occur in the course of the bioethanol production, including cellulase production, cellulose hydrolysis, hexose sugar fermentation, and pentose sugar fermentation (Lynd, 1996). Process configurations proposed for the biological steps differ in the degree to which these steps are integrated (Figure 2). Enzymatic hydrolysis conducted separately from fermentation step is known as separate hydrolysis and fermentation (SHF), which involves four bioreactors. Cellulose hydrolysis performed in the presence of the hexose fermentative microorganisms with cellulase production that is carried out in a separate step by different organisms is referred to as simultaneous saccharification and fermentation (SSF).

When the saccharification of both cellulose (to glucose) and hemicellulose (to xylose, arabinose, galactose, etc.) carried out in SSF with the presence of fermenting microorganisms (usually genetically modified) which can ferment both hexoses and pentoses into ethanol, the SSF would become simultaneous saccharification and co-fermentation (SSCF). Both SSF and SSCF are preferred since both hydrolysis and fermentation can be done in the same tank, resulting in lower costs (Wright et al., 1987). Consolidated bioprocessing (CBP), featuring cellulase production, cellulose and hemicellulose hydrolysis and co-fermentation in one step, is an alternative approach with outstanding potential (Lynd et al., 2005). CBP is the logical endpoint in the evolution of biomass processing technology. Although CBP is less well developed than SSF, it is expected to offer the lowest costs if limitations of current systems can be overcome. The key difference between CBP and other biomass process configurations is that a single microbial community is employed to realize cellulase production, hydrolysis and fermentation simultaneously. Therefore, the benefits associated with CBP would include no capital or operating costs for dedicated enzyme production, reduced diversion of substrate for enzyme production, and compatible systems for enzymatic hydrolysis and fermentation (Lynd, 1996).

The ethanol must be purified for most industrial and fuel uses. After fermentation, ethanol can be recovered from the fermentation broth by distillation or distillation combined with adsorption or filtration, including drying with the addition of lime or a salt, molecular sieves, membranes, addition of an entrainer, and pressure reduction (Gulati et al., 1996; Ladisch et al., 1984; Ladisch and Dyck, 1979; Onuki et al., 2008). The residual solids of distillation process, including lignin, ash, enzyme, organism debris, residue cellulose and hemicellulose, and other components may be recovered as solid fuel or converted to various value-added co-products such as fertilizer and nutrition for sequential fermentations (Hinman et al., 1992; Wooley et al., 1999; Wyman, 1995, 2008).

Although biomass feedstock is available in large quantities, the major challenge for commercialization is to reduce the main operation costs of biomass conversion processes, primarily pretreatment, enzymatic hydrolysis, and fermentation.

BIOMASS PRETREATMENT

Pretreatment is necessary to make cellulose of the lignocellulosic materials more accessible to hydrolytic enzymes by removing the surrounding hemicellulose and/or lignin along with modification of the cellulose microfibril structure. Lots of work has been done on the pretreatment technologies over the past decades. Current research and development activities are focused on the identification, evaluation, development, and demonstration of promising pretreatment techniques which mainly support the subsequent enzymatic hydrolysis of the pretreated biomass (Knauf and Moniruzzaman, 2004). Due to the diverse nature of biomass and different action modes of pretreatment methods, it is difficult or impossible to develop a universal pretreatment technique. An ideal pretreatment process should meet the following requirements (Drapcho et al., 2008):

- High rates of hydrolysis and high yields of fermentable sugars
- Minimal degradation of the carbohydrate fractions
- No production of compounds that are inhibitory to microorganisms used in the subsequent fermentation step
- Inexpensive materials of construction
- Mild process conditions to reduce capital costs
- Recycle of chemicals reduce operating costs
- Minimal wastes

Table 2. Lignocellulosic biomass pretreatment processes

Pretreatment category	Pretreatment method	Reference
Physical pretreatment	Mechanical comminution (milling, attrition, and refining), pyrolysis, irradiation, liquid hot water, uncatalyzed steam explosion (autohydrolysis)	Beardmore et al. (1980); Chandra et al. (2007); Fan et al. (1987); Horton et al. (1980); Rogalinski et al. (2008); Yang et al. (2008); Yoshida et al. (2008); Youssef and Aziz (1999)
Chemical pretreatment	Catalyzed steam explosion, alkaline pretreatment, acid pretreatment, organosolv, ammonia fiber/freeze explosion (AFEX), ammonia recycled percolation (ARP), pH-controlled liquid hot water, wet oxidation, ionic liquids pretreatment	Boussaid et al. (1999); Chang and Holtzapple (2000); Clark and Mackie (1987); De Bari et al. (2007); Fukaya et al. (2008); Herrera et al. (2003); Iyer et al. (1996); Kim et al. (2003); Kim and Holtzapple (2005); Kohlman et al. (1995); Kosan et al. (2008); Mes-Hartree et al. (1988); Murnen et al. (2007); Palonen et al. (2004); Pan et al. (2007, 2008); Prior and Day (2008); Silverstein et al. (2007); Sun and Cheng (2002); Taherzadeh and Karimi (2008); Teixeira et al. (2000); Vazquez et al. (2007); Yoon et al. (1994); Zhu et al. (2004)
Biological pretreatment	Fungi and bacteria	Ander and Eriksson (1978); Hwang et al. (2008); Lee et al. (2007); Schurz (1978)
Combination pretreatment	Two physical pretreatments conducted in sequence, combination of a physical pretreatment and a chemical pretreatment, two chemical pretreatments applied in sequence, a chemical pretreatment followed by a biological pretreatment, a physical pretreatment followed by a biological pretreatment	Chammem et al. (2005); Corredor et al. (2008); Kohlmann et al. (1995); Li et al. (2007); Yu et al. (2009)

Despite no single pretreatment process that can meet all of the above requirements, all the above requirements should be considered in the selection of pretreatment method along with the properties of biomass and characteristics of pretreatment methods. Depending on the application of chemicals (water is not considered as a chemical), pretreatment processes can be grouped into three categories, including physical, chemical and biological pretreatment. Some processes are combinations of two or more pretreatment techniques applied in parallel or in sequence, which are named combination pretreatment. Table 2 summaries the pretreatment processes that have been developed so far.

Most physical pretreatment methods, which are aimed at size reduction and mechanical decrystallization, are limited in their effectiveness and often expensive. Biological pretreatment employs wood degrading microorganisms, primarily including white-, brown-, and soft-rot fungi, and bacteria to modify the chemical composition and/or structure of the lignocellulosic biomass so that the biomass is more amenable to enzyme digestion. The biological pretreatment appears to be a promising technique and has very evident advantages, including no chemical requirement, low energy input, and mild environmental conditions (Hsu, 1996). However, its disadvantages are as evident as its advantages since biological pretreatment is not very effective and requires long treatment time. It also requires careful control of growth conditions and large amount of space to perform treatment (Chandra et al., 2007).

During biological pretreatment, not only lignin but also carbohydrates (hemicellulose and cellulose) are degraded, which usually causes considerable energy content loss (Shi et al., 2008; Singh et al., 2008). Therefore, the studies of biological pretreatment have been limited in bench-scale. It faces techno-economic challenges and is less appealing commercially. The pretreatment which has some chemicals (gases or liquids) involved is defined as chemical pretreatment. It has been originally developed and extensively used in the paper industry for delignification of cellulosic materials to produce high quality paper products. The possibility of developing effective and inexpensive pretreatment techniques by modifying the pulping processes deserves consideration (Fan et al., 1982). Chemical pretreatments that have been explored so far have had the primary goal of improving the enzymatic digestibility of cellulose by removing significant lignin and/or hemicellulose, and to a lesser degree decreasing the DP and crystallinity of the cellulosic component. Chemical pretreatments have gained thus far the most significant attention among all categories of pretreatment techniques. Several chemical pretreatment methods have been tested on pilot and demonstration scales such as catalyzed steam-explosion (Fein et al., 1991; L.O. Ingram, personal communication; Ropare et al., 1992; Tolan, 2006) and dilute sulfuric acid pretreatment (Schell et al., 2003). Combination pretreatment technique incorporates two or more steps consisting over one single pretreatment method from the same or different categories.

Among all the studied combination pretreatment methods, the combinations of physical pretreatment and chemical pretreatment are the most effective combination, and have received the most attention in recent years. Compared to the pretreatment method with a single step, the capital and processing costs of a combination pretreatment method might usually involve additional equipment and/or additional chemicals. The benefits of the combination pretreatment must be balanced against the extra costs.

Table 3. Effect of various pretreatment processes on the compositional and structural features of lignocellulosic biomass (Adapted from Mosier et al., 2005)

Pretreatment method	Increases accessible surface area	Decrystalizes cellulose	Removes hemicellulose	Remove lignin	Alters lignin structure
Uncatalyzed steam explosion	■		■		□
Liquid hot water	■	ND	■		□
pH controlled hot water	■	ND	■		ND
Flow-through liquid hot water	■	ND	■	□	□
Dilute acid	■		■		■
Flow-through acid	■		■	□	■
AFEX	■	■	□	■	■
ARP	■	■	□	■	■
Lime	■	ND	□	■	■

■: Major effect; □: Minor effect; ND: Not determined.

Different pretreatment methods have distinct action mode from each other even though they are classified under the same category. The resulting composition and structure of the pretreated material are dependent on the source of biomass and the type of pretreatment. In general, the pretreatment process makes biomass much more amenable to enzymatic hydrolysis by cellulolytic enzymes than native biomass. The key features of several pretreatment technologies that have potential as cost-effective pretreatments are outlined in Table 3.

A critical limiting factor for all pretreatment processes is their capital-intensive nature. For instance, requirement of costly reactor material, recovery of pretreatment catalysts and waste treatment presents additional costs for overall bioconversion of biomass (Knauf and Moniruzzaman, 2004). The development of advanced pretreatment processes that are tuned to unique characteristics of different types of biomass and minimize cost is still needed along with an improved understanding of pretreatment mechanisms and the relationship between pretreatment and enhancement of enzymatic digestibility of pretreated biomass.

Enzymatic Hydrolysis of Cellulose in the Pretreated Lignocellulosic Biomass

Following the pretreatment, cellulose is more readily to be hydrolyzed into fermentable sugars primarily glucose through either chemical or enzymatic hydrolysis. Although the chemical hydrolysis routes such as acid and alkaline hydrolysis are more established, enzymatic hydrolysis becomes more desirable and offers advantages of high sugar yields, specific reaction, minimal byproduct formation, low energy requirements, mild operating conditions, no corrosion problem, low utility cost, and no special requirements on reactor construction materials (Duff and Murray, 1996; Ghose and Ghosh, 1978; Kadam et al., 1999; Mandels et al., 1974; Van Wyk, 2001). However, it was shown that the cellulolytic enzyme (i.e., cellulase) production was one of the most expensive steps during bioethanol production, even though the enzyme cost has already significantly reduced from 20% to 40% of the total bioethanol production cost (Ruth and Lbsen, 2000; Spano et al., 1978). Significant cost reduction is required in order to enhance the commercial viability of cellulase production technology. During the past decades, lots of work has been done on cellulases, but the costs of these enzymes have remained high, which still hinders the feasibility of commercializing bioethanol as a transportation fuel. To make biomass-based ethanol competitive with gasoline would require further reductions in the current cost of producing cellulase enzymes. Although the current costs of enzymatic route are higher than the established, it has long-term potential for cost reductions (Kadam et al., 2004).

Cellulase Enzymes

A large number of microorganisms, including bacteria and fungi are capable of producing cellulases for cellulose degradation. These microorganisms can be aerobic or anaerobic, mesophilic or thermophilic. There is a distinct difference in cellulolytic strategy between the aerobic and anaerobic microorganisms. Anaerobes degrade cellulose primarily through "complexed" cellulase systems (cellulosome). The cellulosome complex consists of various kinds of enzymes arranged around a scaffolding protein that does not exhibit catalytic activity but enables the complex to adhere to cellulose. The cellulosome allows concerted enzyme activity in close proximity to the bacterial cell wall, enabling optimal synergism between the cellulases existing on the cellulosome. The presence of cellulosome minimizes the distance of cellulose hydrolysis products, allowing efficient uptake of these oligosaccharides by the host cell (Lynd et al., 2002). On the contrary, aerobic cellulose degraders, both bacterial and fungal, hydrolyze cellulose through the actions of substantial amounts of extracellular cellulase enzymes that are freely recoverable from culture supernatants (Lynd et al., 2002; Rapp and Beerman, 1991; Schwarz, 2001). Such anaerobically-produced cellulase systems do not form stable high-molecular weight complexes and therefore are called "noncomplexed" cellulase systems (Lynd et al., 2002). More information about the complexed and noncomplexed cellulase systems including cellulase structure, catalytic mechanisms, cellulase families, and biotechnological applications was provided in a prior review conducted by Lynd et al. (2002).

Table 4. Representative cellulase-producing microorganisms (Adapted from FAO Corporate Document Repository at http://www.fao.org/docrep/w7241e/w7241e08.htm)

Microorganism	Specie
Fungi	*Trichoderma reesei*
	Trichoderma viride
	Thielavia terrestris
	Trichoderma koningii
	Aspergillus acculeatus
	Aspergillus fumigatus
	Aspergillus niger
	Acremonium cellulolyticus
	Talaromyces emersonii
	Fusarium solani
	Irpex lacteus
	Penicillium funmiculosum
	Phanerochaete
	chrysosporium
	Schizophyllum commune
	Sclerotium rolfsii
	Sporotrichum cellulophilum
Bacteria	*Clostridium thermocellum*
	Ruminococcus albus
	Streptomyces sp.
	Bacillus pumilis
Actinomycetes	*Streptomyces sp.*
	Thermoactinomyces sp.
	Thermomonospora curvata

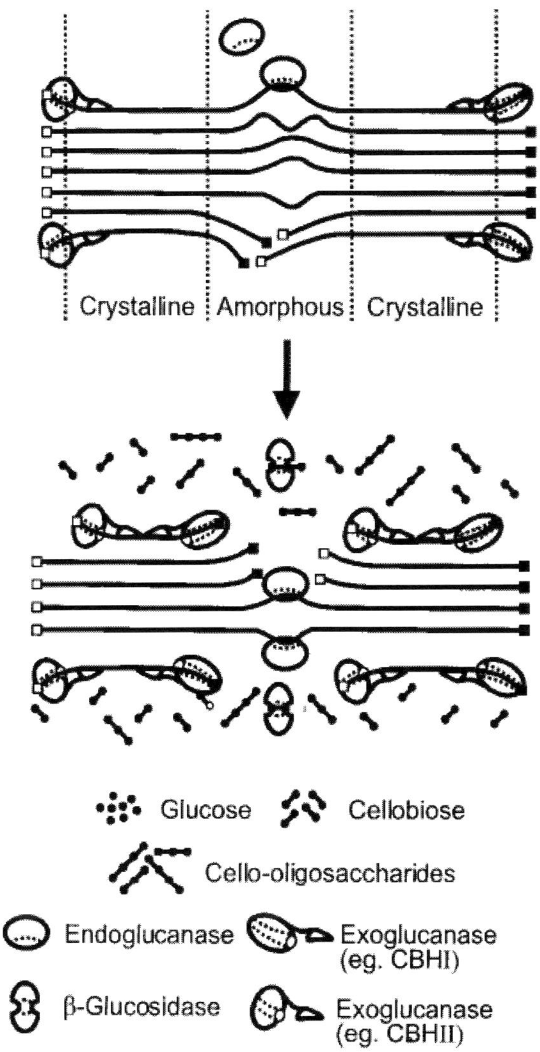

Figure 3. Schematic representation of the cellulose hydrolysis by noncomplexed cellulase systems. (The solid squares represent reducing ends, and the open squares represent nonreducing ends. Cellulose, enzymes, and hydrolytic products are not shown to scale) (Adapted from Lynd et al., 2002).

Although lots of microorganisms can produce cellulase as shown in Table 4, only a few of them produce significant and recoverable quantities of cell-free cellulase enzymes. Anaerobic cellulose-degrading bacteria such as *Clostridium cellulovorans* and *Ruminococcus albus* can produce cellulase with high specific activity; however, they do not produce high enzyme titres (Duff and Murray, 1996; Lynd et al., 2002). In addition, the anaerobes require anaerobic growth conditions and their growth rate is very low. Therefore, cellulases from aerobic fungi have received more research interests, and fungal cellulases currently dominate the industrial applications of cellulases (Gusakov et al., 1992; Nieves et al., 1998). For example, the cellulase system of *Trichoderma reesei* has been extensively researched for more than fifty years (Reese et al., 1950; Reese 1956; Sternberg, 1976).

In particular, fungi of the genera *Trichoderma* and *Aspergillus* are thought to be cellulase producers, and crude enzymes produced by these microorganisms are commercially available for various applications. Either free or cell attached extracellular cellulases must be produced for microorganisms which are capable of hydrolyzing and metabolizing insoluble cellulose. Based on the catalytic action mode and structural properties, the typical cellulase system consists of at least three major components: endo-β-1, 4-glucanase (EG, Cx-cellulase, or EC 3.2.1.4), exo-β-1, 4-glucanase (CBH, cellobiohydrolase, or EC 3.2.1.91), and β-1, 4-glucosidase (cellobiase) (EC 3.2.1.21) (Drapcho et al., 2008; Lee, 1997). The action modes of each of these components shown in Figure 3 include (Coughlan and Ljungdahl, 1988): (1) Cx-cellulase attacks randomly the bonds in the amorphous regions of the cellulose molecules, yielding oligosaccharides of various lengths and consequently new chain ends; (2) Exoglucanase acts in a processive manner on the reducing or nonreducing ends of cellulose chains, releasing either glucose or cellobiose as major products.

Teeri (1997) proposed that exoglucanases can also act on crystalline cellulose, presumably peeling cellulose chains from the crystalline structure; (3) Cellobiase hydrolyzes cellobiose and water-soluble cellodextrins to glucose. For the mechanisms of cellulases' function, Reese et al. (1950) proposed that cellobiohydrolase breaks down the hydrogen bonds in cellulose molecules, followed by hydrolysis of the accessible cellulose with Cx-cellulase. The hypothesis depicted in Figure 3 is apparently the opposite of that proposed by Reese et al. (1950), as indicates that three, rather than two enzymes are essential for the decomposition of cellulose.

Although cellulase isolation and purification techniques have not been fully developed, this hypothesis is now accepted. Therefore, three enzyme groups (EG, CBH and cellobiase) of cellulase systems are not simply agglomerated together but rather act consecutively or in concert to efficiently hydrolyze cellulose (Lynd et al., 2002). The exhibiting collective activities of cellulase systems are higher than the sum of the activities of individual enzymes, which indicates that the individual enzymes hydrolyze cellulose into glucose synergistically (Converse and Optekar, 1993; Medve et al., 1998; Tomme et al., 1990; Väljamäe et al., 1999; Zhang and Lynd, 2004). The forms of synergism could be endo-exo synergy, exo-exo synergy, exo-cellobiase synergy, and/or intramolecular synergy between catalytic domains and carbohydrate-binding modules (CBMs) of cellulase systems (Din et al., 1994; Teeri, 1997).

Industrial fungal cellulases contain some cellobiase activity but at relatively low levels; therefore, a cellobiase such as Novozyme 188, a cellobiase from *Aspergillus niger* is usually needed to supplement the deficit of cellobiase activity to ensure high yields of glucose in the enzymatic hydrolysis of cellulose. In addition to the three major components, commercial cellulase enzymes normally also contain hemicellulase activities such as xylanase, xylosidase, arabinofuranosidase, arabinasase, glucuronidase, acetyl esterase, galactomannanase and glucomannanase (Drapcho et al., 2008; Duff and Murray, 1996).

Since cellulose is an insoluble substrate and cellulases are soluble catalysts, enzymatic hydrolysis is a heterogeneous biological catalytic process. As shown in Figure 4, there are a total of seven steps involved in the enzymatic hydrolysis of cellulosic substrate. Detailed information for every step is described in Table 5.

External diffusion has never been considered as an important factor during enzymatic hydrolysis especially under low substrate loading, because moderate agitation is normally used to mix the mixtures in reactors. Formation of enzyme-cellulose complexes is a prerequisite for cellulose hydrolysis, and such complexes are a central feature of most conceptual and quantitative models for cellulose analysis.

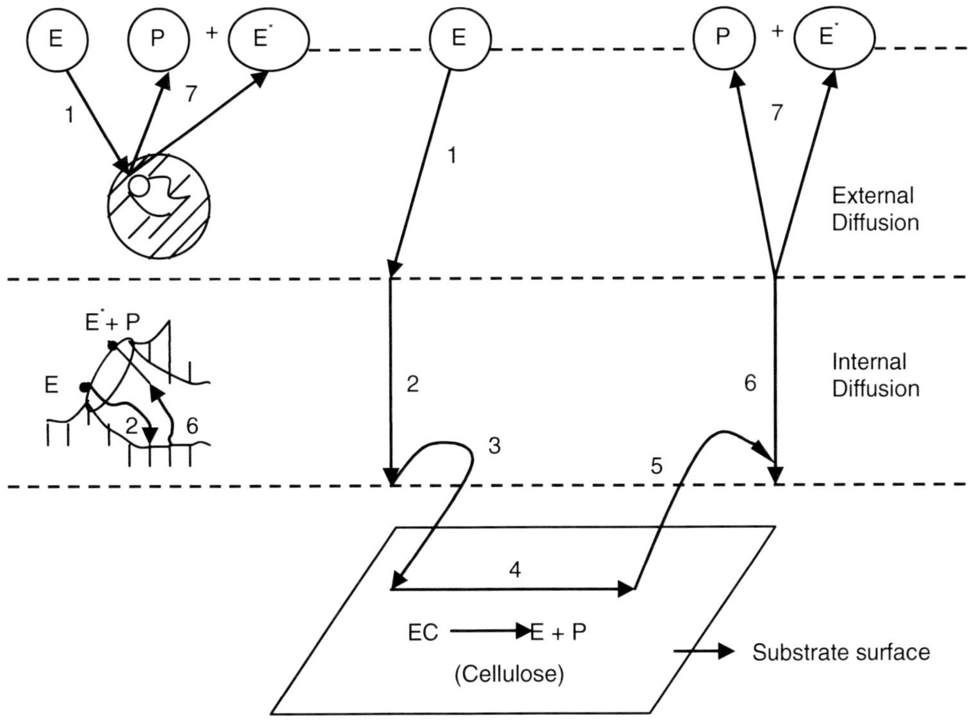

Figure 4. Steps in heterogeneously enzymatic catalytic hydrolysis of cellulose (E are enzymes; P are products; EC are enzymes-products complex; E^* are desorbed enzymes) (Modified based on Fogler, 1986).

Table 5. Steps in enzymatic hydrolysis of cellulosic substrate

Step No.	Descriptions
1	Mass transfer (diffusion) of the enzymes (e.g., E) from the bulk fluid to the external surface of the substrate pellet
2	Diffusion of the enzymes from the pore mouth through the substrate pores to the immediate vicinity of the internal acitive sites of substrate
3	Adsorption of enzymes onto the substrate surface to form enzymes-substrate complex (e.g., EC)
4	Reaction on the surface of substrate (e.g., EC→E+P)
5	Desorption of the products (e.g., P) and/or some of enzymes (e.g., E^*) from the surface
6	Diffusion of products and/or some enzymes from the interior of the pellet to the pore mouth at the extental surface
7	Mass transfer of the products and/or enzymes from the external pellet surface to the bulk solution

In order to understand the enzymatic hydrolysis of the cellulose in pretreated lignocellulosic biomass, it is essential to investigate the adsorption of cellulase enzymes on lignocellulosic biomass components, including cellulose, lignin and whole complex. The adsorption of cellulase enzymes onto pure cellulose and various types of lignocellulosic biomass has been extensively studied in order to indentify the parameters controlling the rate of cellulose hydrolysis and to design with this knowledge more efficient conditions for the practical application of cellulase enzymes in the conversion of lignocellulosic biomass to fermentable sugars. Prior reviews consider the cellulase adsorption phenomena and quantification of hydrolysis (Lynd et al., 2002; Zhang and Lynd, 2004).

Enzymatic Hydrolysis Process

Cellulases are relatively slow-acting enzymes, primarily due to the complex nature of cellulosic substrates and requirements of synergistic actions between multiple enzyme groups of cellulases on the hydrolysis of cellulosic substrate. Therefore, not only do advances in enzyme development have to achieve classical enzyme improvements such as stability, yield, and increased specific activity but they also need to be effective in harsh environments generated by the pretreatment process (Knauf and Moniruzzaman, 2004). In addition to the engineering of improved cellulase enzymes, the factors that affect the enzymatic hydrolysis including substrates, cellulase activity, end-product inhibition, and reaction conditions (temperature, pH, as well as other parameters) have been studied with the focus on the optimization of the hydrolysis process in order to realize high yield and rate of the enzymatic hydrolysis with relatively low enzyme dosage.

The main effects of substrate on the yield and rate of enzymatic hydrolysis of cellulose include the substrate concentration and its structural features. An increase of substrate concentration normally leads to an increase of the yield and reaction rate of the hydrolysis (Cheung and Anderson, 1997). However, substrate concentration can cause substrate inhibition when it exceeds a critical point, which substantially lowers the rate of the hydrolysis (Tengborg et al., 2001; Zheng et al., 2009a). The extent of substrate inhibition was found to depend on the ratio of total substrate to total enzyme (Huang and Penner, 1991; Penner and Liaw, 1994). In addition, the heat and mass transfer problems associated with high substrate concentration also result in low yield and rate of enzymatic hydrolysis. Fed-batch succeeded in achieving higher yield and rate of enzymatic hydrolysis than batch under high substrate concentration (Zheng et al., 2009a). In fed-batch hydrolysis, substrate and/or enzymes are added into reactors stepwise and substrate are gradually liquefied and degraded; therefore, the mixture becomes more fluid and higher substrate concentration could be added without the occurrence of substrate inhibition (Rudolf et al., 2005). Fed-batch is expected to be a better procedure than batch on dealing with the situation of high substrate concentration with relatively low enzyme dosage while greatly saving the capital cost of reactors. Additionally, fed-batch can generate high glucose concentration, which is translated into high ethanol concentration making ethanol recovery more economical by distillation (Ballesteros et al., 2002).

In addition to the concentration, the degradability of cellulosic substrate also depends on the structural features including the extent of crystallinity, degree of polymerization, deacetylation, accessible surface area, as well as the presence of lignin. Based on the review of the previous research results, contradictory conclusions were found on the effects of crystallinity and degree of polymerization on enzymatic hydrolysis (Lynd et al., 2002; Zhang and Lynd, 2004), but consistent on the behavior of deacetylation, accessible surface area and lignin content (Hsu, 1996). Lignin has negative impact on the enzymatic hydrolysis by blocking access of cellulases to cellulose and by irreversibly adsorbing cellulase enzymes nonproductively; therefore, lignin removal is very important to increase the hydrolysis rate (McMillan, 1994; Zheng et al., 2008, 2009b). The addition of certain types of surfactants and proteins can reduce or eliminate the detrimental effect of lignin on the enzymatic hydrolysis of cellulosic substrate (Eriksson et al., 2002; Helle et al., 1993; Kaya et al., 1995; Ooshima et al., 1986; Zheng et al., 2008). Since different pretreatment method processes biomass in a unique way, yielding pretreated substrate with different composition and structure. As a result, cellulase enzymes have to be tailored to the unique composition and structure of the pretreatment process, which means the hydrolytic enzyme systems need to be optimized based on specific pretreated cellulosic substrate. For example, the dilute acid pretreated biomass usually needs cellulase and supplemented cellobiase; however, lime pretreated biomass might need extra hemicellulase in addition to cellulase and cellobiase.

Cellulase loading is one of important factors affecting the yield and rate of enzymatic hydrolysis. Due to the complex nature of cellulosic substrate and end-product inhibition, high cellulase dosage is usually required to enhance the yield and rate of hydrolysis within elongated residence time during enzymatic hydrolysis (Tengborg et al., 2001; Zheng et al., 2009a). Increases in the cellulase concentration and long reaction time would significantly increase the final ethanol cost. The cellulase loading varied from 5 to 70 FPU/g-cellulose within 2-7 days of hydrolysis, depending on the type and concentration of cellulosic substrate. Cellulase dosage of 10 FPU/g-cellulose is often used in laboratory and technoeconomic modeling studies because it provides a hydrolysis profile with high levels of glucose yield in a reasonable time (2-3 days) at a reasonable enzyme cost (Gregg and Saddler, 1996). Therefore, the yield and rate of hydrolysis must be balanced against the cost of cellulase enzymes.

Since it is too costly to improve enzymatic hydrolysis by increasing enzyme loading, other attempts have been tried and met with some success. It is well known that end-product inhibition of cellulase results in severe cellulase activity loss during hydrolysis. The patterns of these inhibitions by glucose and cellobiose have been the subject of research for a long time, with many differing views on the nature of this inhibition (Holtzapple et al., 1990; Zhang and Lynd, 2004). Cellobiose was believed to be stronger inhibitor of cellulase than glucose while the supplemented cellobiase is more sensitive to glucose inhibition. Several methods have been used to mitigate the inhibition, including the use of high concentration of cellulase (Dekker et al., 1987; Ishihara, et al., 1991), the supplementation of cellulase with cellobiase during hydrolysis (Breuil et al., 1990, 1992; Excoffier et al., 1991; Mandels et al., 1981; Spindler et al., 1989a; Tan, et al., 1987; Wright et al., 1986; Xin et al, 1993), and the removal of product sugars during hydrolysis by ultrafiltration (UF) (Ding et al., 1998; Ishihara, et al., 1991; Tan et al., 1986, 1987) or simultaneous saccharification and fermentation (SSF) (Mes-Hartree et al., 1987; Philippidis et al., 1992; Saddler, et al., 1982; Spindler et al., 1989a; Szczodrak and Targonski, 1989). The former two methods are undesirable due to the cost issues; therefore, UF and SSF have been two main approaches for alleviating the end-product inhibition. Although UF can help remove the end-products, deprived efficiency was obtained because of the Cx-cellulase leaching from the system and subsequently reducing the synergistic interaction between cellulase enzymes required for the efficient hydrolysis of cellulose (Tan et al., 1987). Besides, the final sugar concentration of the UF equipped hydrolysis system is too low to achieve recoverable ethanol after fermentation (Ding et al., 1998). In the SSF process, the sugars' inhibitions are prevented via continuous microbial fermentation of glucose to ethanol as soon as glucose is produced (Philippidis et al., 1993). In addition to alleviating end-production inhibition, SSF has several advantages over a separated hydrolysis and fermentation process (SHF) of higher hydrolysis rate, higher ethanol yield and concentration, lower required enzyme loading, reduced costs of reaction and separation equipments, shorter process time, and lower requirements for sterile conditions because of shorter retention time of glucose. Although sugar inhibition is significantly reduced, ethanol exhibits substantial inhibition to the cellulase activity in the SSF process (Philippidis et al., 1992; Wu and Lee, 1997). It was found that cellulase lost its original activity of 9%, 36% and 64% with the presence of ethanol of 9, 35 and 60 g/l, respectively (Wu and Lee, 1997). In SSF, the enzymatic hydrolysis has to run at a compromised temperature of around 30-37°C due to the temperature sensitivity of the fermenting microorganisms such as *Saccharomyces cerevisiae* and Escherichia coli KO11, instead of the enzyme optimum temperature of 45-50°C. Another major disadvantage associated with SSF lies in the difficulty of separating the cells and enzymes from the unhydrolyzed solid residue, as restricts the reuse of the high concentrations of yeasts that are necessary to ensure good ethanol production in the subsequent SSF (Eklund, 1994; Gregg and Saddler, 1996). Development of high ethanol tolerant fermenting microbes is also needed to be considered in SSF. In addition to the above methods for reducing the end-production inhibition, some researcher have proposed another two methods such as genetic modification to make *T. reesei* cellulases contain high β-glucosidase activity (Tshiteya, 1992) and addition of cellobiose-fermenting yeast into SSF process (Wyman et al., 1986; Spindler, et al., 1988; 1989b; 1992). However, these two methods have not achieved significant success so far.

To improve enzymatic hydrolysis, cellulase mixture from different cellulase producers or a mixture of cellulases and other enzymes have been used. The most common example is the addition of cellobiase (Novozyme 188) into the *T. reesei* cellulases system (Celluclast) achieved better saccharification than the system without cellobiase addition (Zheng, 2007; Zheng et al., 2009a). Cellulase has also been mixed with xylanase or pectinase to achieve significant increase of cellulose conversion (Beldman et al., 1984; Ghose and Bisaria, 1979; Saha et al., 2008). In addition, cellulase mixture from different microorganism has been used to hydrolyze alkali-treated sugarcane bagasse (Mononmani and Sreekantiah, 1987), steam-explosion pretreated Eucalyptus chips (Ramos et al., 1993), and microcrystalline cellulose (Baker et al., 1994). For example, Ramos et al. (1993) obtained almost 100% of cellulose conversion with substrate concentration of 6% and enzyme loading of 10 FPU (cellulase mixture)/g-cellulose.

During enzymatic hydrolysis, the irreversible adsorption of cellulase on unhydrolyzed cellulose and lignin content is partially responsible for enzyme deactivation (Converse et al., 1988; Zheng et al., 2008). Surfactants (e.g. Tween 20, Tween 80, Triton-X100, etc.) and non-catalytic proteins (e.g. Bovine Serum Albumin-BSA and gelatin) have been used to enhance the yield and rate of enzymatic hydrolysis by modifying the surface property of cellulosic substrate and minimizing the irreversible binding of cellulase on substrate. The enzyme loading can be saved up to 50% while maintaining similar glucose yield with the presence of Tween 20 or BSA in the hydrolysis of dilute acid pretreated Creeping Wild Ryegrass (Zheng et al., 2008). The rate of enzymatic hydrolysis was improved by 33% using Tween 80 as a surfactant in the hydrolysis of newspaper (Castanon and Wilke, 1981).

Reaction parameters such as temperature and pH also play important roles in the enzymatic hydrolysis of cellulosic biomass. For most fungal cellulase enzymes, maximum cellulase activity is observed at 50± 5 °C and a pH of 4.0-5.0 (Saddler and Gregg, 1998). According to the manufacturer, optimum conditions for cellulase complex are temperature around 45-50°C and pH values between 4.5 and 6.5 (product sheet, Novozymes). Reviewing the previous research on the enzymatic hydrolysis of various cellulosic substrates for enzymes obtained from *Trichoderma* and *Aspergillus*, Tengborg et al. (2001) summarize that optimal temperature and pH for these fugal cellulases are in the range from 40 to 50 °C, and from 3.8 to 5.0, respectively. As noted that most previous research optimized hydrolysis temperature (or pH) under single pH (or temperature) value. They did not study the effect of interaction between pH and temperature or other physical parameters on the determination of optimal parameters. Using response surface methodology, Tengborg et al. (2001) found that the optimal temperature is affected not only by the pH but also by the residence time when studying the enzymatic hydrolysis of fresh chipped spruce (*Picea abies*) pretreated by SO_2 catalyzed steam-explosion (Figure 5). Therefore, it is of importance to investigate both individual parameter effect and interaction effect between different parameters to determine the maximum cellulose hydrolysis yield.

Taking into account interaction nature between cellulase and cellulosic substrate, cellulase enzymes can be recovered from either the liquid supernatant or the solid substrate.

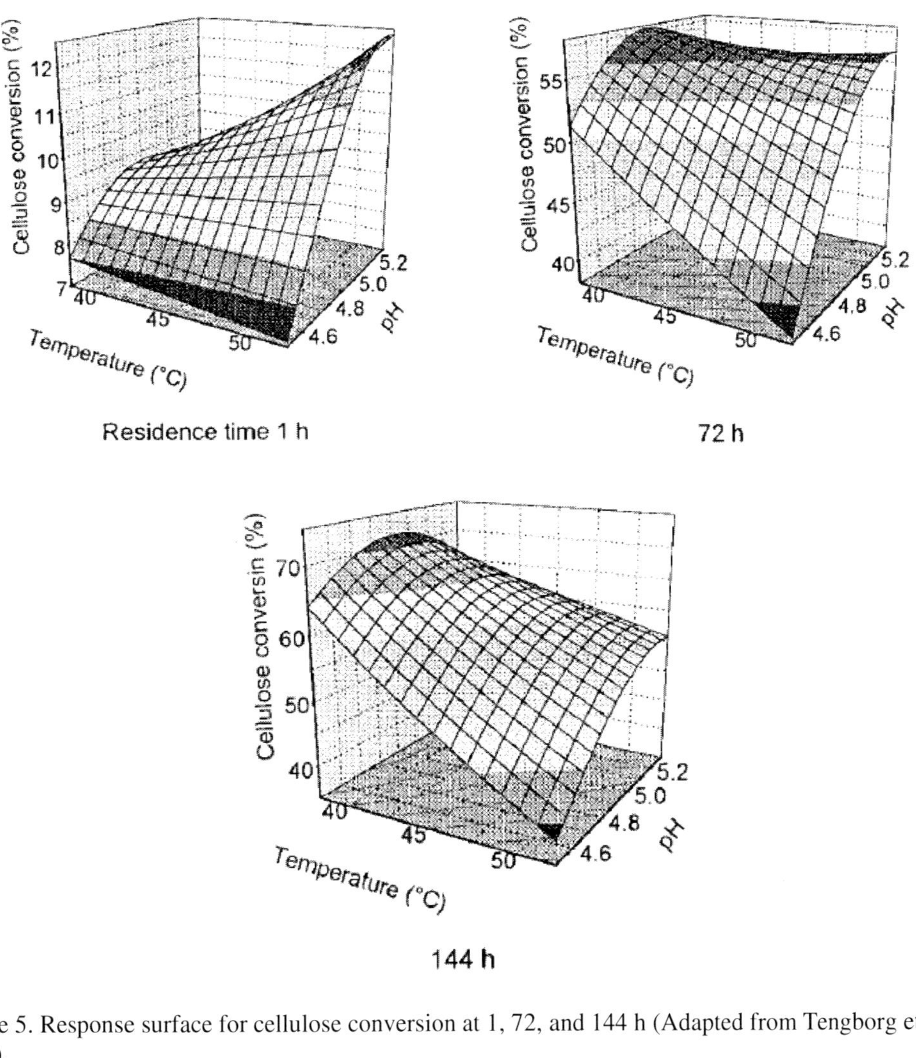

Figure 5. Response surface for cellulose conversion at 1, 72, and 144 h (Adapted from Tengborg et al., 2001).

Therefore, three possible enzyme recovery scenarios could be developed including recovery of the enzymes associated with the substrate, enzymes in the supernatant and associated with the substrate, and enzymes free in the supernatant (Gregg and Saddler, 1996). Most enzyme recycling techniques so far deal with enzymes in the supernatant such as the addition of fresh substrate for reuse of enzymes (Singh et al., 1991) and ultrafiltration (Ramos and Saddler, 1994).

When lignin content of substrate is relatively low, the method by using direct readsorption of the enzyme onto fresh substrate achieved some successes (Girard and Converse, 1993), but the recovery of enzyme activity varies inversely with the lignin content of the substrate. For example, the recovered cellulase activity from high lignin containing substrate such as steam-exploded wheat straw (23% lignin content) and aspen (21% lignin content) achieved only 32-55% of the original sugar yields (Vallander and Eriksson, 1987). However, Ramos et al. (1993) obtained almost 100% and 95% of the original sugar yields after one and five rounds, respectively when the cellulases associated with steam-exploded, peroxide-treated eucalyptus (0.7% lignin content) was recycled and reused for hydrolysis of the pretreated eucalyptus. The regular means to recover the enzymes associated with substrate consisting of relatively high lignin is buffer elution, which employ a large amount of specific buffers and/or detergents to gradually wash the substrate. It is, however, faced with significant technical and economic difficulties associated with the scale-up because of utilization of costly chemicals and the very dilute enzyme in the large amount of washing buffer solution. To reduce the detrimental effect of lignin on cellulase recycling, surfactants have been extensively studied to improve the cellulase recycling by preventing the irreversible and nonproductive adsorption of cellulase on lignin. Castanon and Wilke (1981) achieved 3-fold increase of enzyme activity recovery when hydrolyses were carried out on the newspaper in the presence of Tween 80. Tu et al. (2007) also found that the nonionic surfactants Triton X-100, Tween 80, and Tween 20 enhanced a single round of cellulase recycling performance (based on the sugar yield) for the ethanol pretreated Lodgepole pine (EPLP) substrate by 50%. Since the commercial price of surfactants is not available, the economic feasibility of surfactant utilization is not clear. To find cost-effective enzyme recovery strategies, the future research could aim at the development of novel pretreatment methods which can remove most lignin content from cellulosic substrate, make pretreated lignin low affinity to the cellulases, or produce highly enzymatically digestible pretreated substrate. As summarized by Gregg and Saddler (1996), enzyme recycle has several economic benefits, including enhanced rates and yields of hydrolysis, lower enzyme costs, and decreased capital costs due to shorter residence times. Thus, more studies need to be conducted in future to further improve the economic feasibility of enzymatic saccharification of cellulosic biomass for ethanol production.

FERMENTATION OF SUGARS INTO ETHANOL

Hydrolysis of lignocellulosic biomass yields a mixture of sugars, which include both hexoses (glucose, galactose, and mannose) and pentoses (xylose and arabinose), with glucose, xylose, and/or arabinose being the most predominant ones. Various microorganisms, either yeasts, bacteria, or fungi have been reportedly used to ferment these carbohydrates to ethanol with the absence of oxygen. The net fermentation result can be expressed as:

Carbohydrate (CH_2O) + Cells → Ethanol (C_2H_5OH) + CO_2 + More cells

From the above equation, the theoretical ethanol yield from carbohydrate is calculated to be 0.51 g ethanol/g carbohydrate without the consideration of cell mass. This theoretical yield can never be achieved in practice since not all of the carbohydrate consumed is converted to ethanol but part of it (5-12% of consumed carbohydrates) is used to support cell growth, cell maintenance, and by-product production. Therefore, less than 47% carbohydrate is usually converted into ethanol within most proposed ethanol production processes (Lynd, 1996). Under ideal conditions, 90-95% of theoretical yield can be realized (Thomas et al., 1996).

Not all naturally occurring ethanol producing microorganisms can use both hexoses and pentoses for ethanol production. Hexoses such as glucose derived from cellulose are readily fermented into ethanol by all ethanologenic microorganisms. On the contrary, most native microorganisms are unable to digest pentoses such as xylose and arabinose into ethanol. For example, *S. cerevisiae*, which has been extensively used for industrial ethanol production from sugar and starch feedstocks, can't metabolize pentoses to ethanol. Therefore, *S. cerevisiae* does not fully satisfy the ethanol production from lignocellulosic biomass. Since pentoses resulted from hemicellulose usually account for 20-35% of lignocellulosic biomass, commercially-interested cellulosic ethanol must necessitate efficient utilization of pentoses except for hexoses.

The organism or system of organisms, which are ideally suitable for industrial cellulosic ethanol production in a process featuring enzymatically mediated hydrolysis, should possess the following traits (Dien et al., 2003; Lynd, 1996; Drapcho et al., 2008):

- Simultaneous ethanol fermentation and synthesis of an active cellulase enzyme system at high levels without addition of extra costly cellulase enzymes
- Conversion of a wide range of biomass-derived sugars (hexoses and pentoses) to ethanol
- High growth rate and high ethanol yield (>90% of theoretical yield)
- Production of ethanol at high concentration and selectivity
- Tolerance of high ethanol level (>40 g/L)
- High ethanol productivity (>1 g/L/h)
- Resistance to inhibitors generated during the biomass pretreatment
- Robust grower and no requirements for complicated and expensive medium formulation
- Culture growth conditions retard contaminants, that is, low pH and high temperatures

Mesophilic Ethanol Producing Microorganisms

Historically, the yeast has been the most commonly used organism for ethanol production. Among the yeast, *S. cerevisiae* is the most preferred one for most ethanol fermentation processes. It is also generally recognized as safe (GRAS) as a food additive for human consumption (Lin and Tanaka, 2006). It can metabolize various sugars, including glucose, fructose, mannose, galactose, sucrose, maltose, and/or maltotriose into ethanol via Emben-Myerhof-Parnas (EMP) pathway. Under optimal conditions, *S. cerevisiae* can produce up to 18%, by volume, ethanol with 15 to 16% being the norm. Wild *S. cerevisiae*, however, was unable to ferment xylose or arabinose into ethanol. Thus, many efforts have been made to successfully create recombinant *S. cerevisiae* capable of utilizing xylose and arabinose by either transforming the host with plasmids harboring the genes required for xylose or arabinose utilization for ethanol production or integrating the genes required for xylose or arabinose metabolism into the chromosome of the host (Brat et al., 2009; Ho et al., 1998; Kuyper et al., 2004, 2005; Op Den Camp et al., 2003; Runquist et al., 2009; Wisselink et al., 2007, 2009). For example, strain *S. cerevisiae* 424A (LNH-ST) was able to co-ferment glucose and xylose to ethanol with yield ranging from 75-97% of theoretical value depending on sugar resources (Sedlak and Ho, 2004); and strain *S. cerevisiae* IMS0002 exhibited high rates of arabinose consumption (0.70 g dry cell weight/h) and ethanol production (0.29 g/g dry cell weight/h) and a high ethanol yield (84% theoretical yield) during anaerobic growth on L-arabinose as the sole carbon source (Wisselink et al., 2007). Recently, Wisselink et al. (2009) have developed a novel strain *S. cerevisiae* IMS0003 capable of fermenting mixtures of glucose, xylose, and arabinose with a high ethanol yield (0.43 g/g total sugar) without formation of the side products xylitol and arabinitol.

Except for the yeast, various ethanologenic bacteria, especially including *Zymomonas mobilis*, *Escherichia coli*, and *Klebsiella oxytoca*, have also attracted lots of research interests. And they currently show the most promise for industrial exploitation. *Z. mobilis* is considered the most effective ethanol producing organism even though it is not currently employed for commercial ethanol production. It is a really unique Gram-negative bacterium that metabolizes glucose, sucrose and fructose anaerobically to pyruvate via the Entner-Doudoroff (ED) pathway, which is used mostly by strictly aerobic organisms. The ED pathway produces only half as much ATP per mole of glucose as the EMP pathway used by *S. cerevisiae* (Dien et al., 2003). Thus, the generated cell biomass of *Z. mobilis* during fermentation is less than the yeast, which means more carbohydrates are funneled to fermentation products through the ED pathway. In other words, *Z. mobilis* cell growth and high biomass concentration are not necessary for ethanol production and high ethanol yield, respectively. As a consequence, using *Z. mobilis* in the ethanol fermentation processes could achieve high ethanol yield close to the theoretical value. In general, *Z. mobilis* has several advantages over *S. cerevisiae* of high sugar uptake rate (Gunasekaran and Raj, 1999; Jeffries, 2005), high specific ethanol productivity (2.5 times higher than *S. cerevisiae*) (Rogers et al., 1982; Sprenger, 1996), high ethanol tolerance (~120 g/L ethanol) (Dien et al., 2003; Lin and Tanaka, 2006), high ethanol yield (~97% of the theoretical yield) (Jcffries, 2005), no requirements for aeration (Gunasekaran and Raj, 1999; Jeffries, 2005), and simple nutritional needs (Dien et al., 2003; Rogers et al., 1982). Additionally, *Z. mobilis* is also GRAS. Such advantages of *Z. mobilis* are consequences of its unique physiology, i.e. the ED pathway. However, *Z. mobilis* is not well suited for industrial ethanol production in spite of its capability of producing ethanol from glucose with the above advantages. The disadvantages that prevent *Z. mobilis* from being used for commercial production of ethanol include low ethanol yield upon sucrose due to by-product production, requirements for sterilization of fermentation medium, strict control and maintenance of culture/inoculum needed, and no fermentation of pentoses such as xylose and arabinose. Although *Z. mobilis* is not used commercially for ethanol production, research interest in this organism remains high. Optimization of bioreactor system such as immobilized *Z. mobilis* cell with enzyme biocatalysts in fluidized-bed reactor has been conducted by Oak Ridge National Laboratory to improve the ethanol production (Nghiem and Davison, 2002). The development of recombinant *Z. mobilis* strains with broader spectrum of sugars (e.g. glucose, xylose and arabinose) utilization for ethanol production is currently the primary research focus, as was well reviewed by Dien et al. (2003), Drapcho et al. (2008), Lin and Tanaka (2006), and Lynd et al. (2002).

Since native *E. coli* is already an ethanol producer and can utilize both hexose and pentose sugars and the uronic acid constituents of pectin, it is a good candidate for cellulosic ethanol production and has attracted lots of research attention. Additionally, the growth of *E. coli* does not require complex medium and it has been extensively used to produce industrial recombinant protein, as further increases its potential for ethanol production. The major disadvantage of *E. coli* is the ethanol yield is fairly low, only 25% of the theoretical value of 0.51 g ethanol/g glucose (Gottschalk, 1986). Unlike the yeast and *Z. mobilis*, *E. coli* is hetero-ethanol fermentative and usually metabolizes sugars into not only ethanol but also a number of by-products, primarily including succinatic, lactatic and acetic acids. Additional disadvantages associated with *E. coli* utilization for ethanol production includes near neutral and narrow pH tolerance range (ca. pH 6.0-8.0), less hardiness compared to the yeast, public safety concern, and lack of data on the application of *E. coli* fermentation residuals.

The breadth of carbohydrates metabolized, extensive background of knowledge, and ease of genetic manipulation made *E. coli* an obvious choice for metabolic engineering of a microbial biocatalyst for production of ethanol from lignocellulose. Thus, metabolic engineering has been successfully applied to construct recombinant *E. coli* strains to selectively produce ethanol on various sugars including glucose, xylose and arabinose from the 1980s in order to improve the ethanol production effectiveness using *E. coli*. Several promising strains include *E. coli* ATCC11303 (pLOI297) (Ingram et al., 1987), *E. coli* KO11 (Ohta et al., 1991a), *E. coli* FBR5 (Dien et al., 2000), *E. coli* LY01 (Yomano et al., 1998), and *E. coli* LY168 (Yomano et al., 2008). Recently, Ingram's Laboratory has successfully developed a new *E. coli* strain with improved inhibitor tolerance and inexpensive culture medium (L. O. Ingram, personal communication). Refer to literatures for more details about the development of ethanologenic *E. coli* strains (Dien et al., 2003; Ingram et al., 1999; Jarboe et al., 2007).

The other extensively-studied ethanologenic bacterium is *Klebsiella oxytoca*, which is of particular interest because it can metabolize a variety of hexoses and pentoses, as well as cellobiose and cellotriose. This trait may reduce or eliminate the requirement of the costly β-glucosidase which normally is used to supplement commercial cellulases in cellulose hydrolysis. It may also require less dosage of cellulase during cellulose hydrolysis and fermentation compared with non-cellobiose fermenting organisms by means of alleviating the inhibition of cellobiose on cellulases (Freer and Detroy, 1983). Another potential advantage associated with this trait would be the decrease of some contaminants via the reduction of glucose in the fermentation broth (Dien et al., 2003). Overall, this trait makes *K. oxytoca* especially attractive for cellulose fermentation for ethanol production. *K. oxytoca* ferments sugars to neutral products such as butanediol and ethanol, as well as a variety of organic acids, including acetic, lactic and succinic acids (Ohta et al., 1991b). In order to develop recombinant *K. oxytoca* strains with selective ethanol production, Ohta et al. (1991b) introduced the production of ethanol (PET) operon into *K. oxytoca* to create several plasmid -

based strains using *K. oxytoca* strain M5A1, including M5A1 (pLOI297), M5A1 (pLOI555) and M5A (pLOI560) and found that M5A1 (pLOI555) achieved ethanol yields up to 98% and 94% of theoretical yield upon glucose and xylose, respectively. By using these plasmid-based strains, the ethanol selectivity was increased to 90% and the productivity reached 2 g/L/h during the first 24 h. Wood and Ingram (1992) further transformed the *K. oxytoca* M5A1 into integrated strains by inserting PET operon into its chromosome. They used the method, which was used for the selection of hyper ethanol-producing *E. coli* transformant, to select recombinant *K. oxytoca*. The best mutant strain P2 readily converted 100 g/L cellobiose or glucose into 45 g/L ethanol (88% theoretical yield) in 48 h without additional β-glucosidase. *K. oxytoca* P2 has been successfully tested on mixed sugars (L-arabinose, D-xylose and D-glucose) (Bothast et al., 1994), mixed office paper (Brooks and Ingram, 1995), actually hydrolysates of sugarcane bagasse by diluate acid and AFEX (Doran et al., 1994), corn fiber (Moniruzzaman et al., 1996), and sugar beet pulp (Doran et al., 2000).

Although the recombinant *S. cerevisiae*, *Z. mobilis*, *E. coli*, and *K. oxytoca* have been created and optimized to selectively produce ethanol with relatively high yield from a broad range of sugars, they can't consume cellulose or xylan directly to produce ethanol. As a result, extra cellulase, β-glucosidase, xylanase, and/or β-xylosidase are usually needed in SSF process to hydrolysis cellulose and/or xylan into sugars, which are sequentially fermented by these organisms into ethanol. More recently, attempts have been made to create cellulolytic and xylanolytic strains with the capability of fermenting cellulose and/or xylan. Fujita et al. (2004) developed a cellulolytic *S. cerevisiae* by coexpressing endoglucanase II (EG II) and cellobiohydrolase II (CBH II) derived from *T. reesei* and β-glucosidase I of *A. aculeatus* on the cell surface of the yeast. By using this modified yeast strain, 89% of theoretical ethanol yield was achieved within 40 h upon 10 g/L amorphous cellulose. Similarly, Fujita and his coworkers created a xylanolytic yeast by using a cell surface engineering system based on α-agglutinin to construct a *S. cerevisiae* strain codisplaying two types of xylan-degrading enzymes, namely, xylanase II (XYNII) from *T. reesei* QM9414 and β-xylosidase (XylA) from *A. oryzae* NiaD300, on the cell surface. The modified strain was able to hydrolyze xylan to xylose (Katahira et al., 2004). This xylan-utilizing *S. cerevisiae* strain was then transformed with plasmids harboring genes for xylose utilization, namely, those encoding xylose reductase and xylitol dehydrogenase from *Pichia stipitis* and xylulokinase from *S. cerevisiae*. The transformed strain produced 7.1 g/L (ca. ethanol yield of 0.3 g/g xylan consumed) ethanol directly from birchwood xylan after 62 h in a SSF process. Similar work has been done to construct a cellulolytic *K. oxytoca* strain. Genes (CelZ and CelY) encoding extracellular endoglucanase activities from *Erwinia chrysanthemi* were integrated into the chromosomal DNA of *K. oxytoca* strain P2 to create a cellulolytic *K. oxytoca* strain SZ21 using a surrogate promoter from *Z. mobilis* for expression (Zhou et al., 2001). The endoglucanase activity produced by strain SZ21 was so low that a certain amount of commercial cellulase is still needed for efficient cellulose (Sigmacell 50) conversion. With the presence of additional exogenous cellulase, strain SZ21 was able to yield up to 22% more ethanol compared with the parental strain. Except for the engineering cellulolytic and xylanolytic strains, a wild obligately anaerobic, mesophilic, cellulolytic bacterium, *Clostridium phytofermentans* strain $ISDg^T$, was isolated from forest soil. The optimum temperature for growth is 35-37 °C. It was found that a wide range of carbohydrates, including cellulose, pectin, polygalacturonic acid, starch, xylan, arabinose, cellobiose, fructose, galactose, gentiobiose, glucose, lactose, maltose, mannose, ribose and xylose, support the growth of *C. phytofermentans*. The major end products of fermentation include ethanol, acetate, CO_2 and H_2; formate and lactate were minor products. *C. phytofermentans* is of particular interest for the production of high concentrations of ethanol during cellulose fermentation. Two to four times more ethanol than acetate are generated, suggesting that *C. phytofermentans* possesses unusual fermentation pathways (Warnick et al., 2002).

The development of cellulolytic and xylanolytic ethanol-producing organisms discussed above is one strategy to decrease the cellulosic ethanol production costs by reducing or eliminating the addition of expensive exogenous cellulase and β-glucosidase during ethanol production processes, e.g. SSF, SSCF and/or CBP. The cellulolytic organisms, which have the capability of fermenting a broad spectrum of sugars such as *C. phytofermentans*, would be most suitable for CBP process.

Thermotolerant Ethanol Producing Microorganisms

The ethanol-producing yeast and bacteria discussed above, including *S. cerevisiae*, *Z. mobilis*, *E. coli*, *K. oxytoca*, and *C. phytofermentans* are all mesophilic. Their optimal growth temperature is 30-37 °C. The optimal pH for most of them is 4.5-5.0, but *E. coli* has near neutral and narrow range of tolerable pH (ca pH 6-8). Commercial cellulases such as *T. reesei* cellulase usually have optimal activity at pH 4.5-5.0 and 50±5 °C. However, SSF operation requires that enzyme and culture conditions be compatible with respect to pH and temperature. Thus, the recombinant *E. coli* is the least suitable organism for SSF. As a consequence, a compromise of the optimal temperatures and/or pH for the cellulolytic enzymes and the fermenting organisms is needed. For instance, SSF experiments were often conducted at 37°C and pH 4.5-5.0 when *S. cerevisiae* was used for fermentation. This was regarded as a suitable compromise at the high end of what *S. cerevisiae* can tolerate.

Therefore, thermotolerance is clearly an important topic for SSF and much work has been directed at developing thermotolerant and thermophilic organisms for ethanol fermentation. Thermophilic ethanol fermentation refers to the fermentation at high temperature (ca 60-70 °C). In addition to the common characteristics of broad spectrum of sugar utilization, thermotolerant and thermophilic ethanologenic organisms offer a number of advantages, including:

- Reduction or elimination of the need for power-consuming cooling and aeration of large reactor vessels
- Increase of the rate of ethanol productivity
- Elimination of contamination problem due to high fermentation temperature so as to reduce or eliminate the need of costly sterilization process
- Continuous removal of volatile ethanol by evaporation and distillation under mild vacuum to lower the ethanol recovery cost
- Reduction of ethanol inhibition due to low ethanol concentration in fermentation broth by the constant removal of ethanol
- No requirements of an organism that is tolerant to very high ethanol concentration

Various thermophilic ethanologenic bacteria have been used for ethanol fermentation, such as *Clostridium thermocellum* (Lynd et al., 1989; NG et al., 1981; Rani et al., 1998; Shama, 1991; Shin et al., 2002), *Clostridium thermohydrosulfuricum* (Germain et al., 1986; NG et al., 1981), *Clostridium thermosaccharolyticum* (Saddler and Chan, 1984; Venkateswaren and Demain, 1986; Wang et al., 1983), *Thermoanaerobacter ethanolicus* (Wiegel and Ljungdahl, 1981), *Thermoanaerobium brockii* (Lamed and Zeikus, 1980), *Thermoanaerobacter mathranii* (Ahring et al., 1996; Larsen et al., 1997), and *Bacillus stearothermophilus* (Hartley, 1993; Hartley and Shama, 1987; Martin et al., 1993; Shama, 1991). Among all these thermophilic ethanol producing organism, *Clostridium thermocellum* received the most intensive research due to its advantages on ethanol fermentation as shown in Table 6. It was the first thermophilic organism isolated from anaerobic digesters, thermal springs, and stacked manure by Viljoen et al. (1926). *C. thermocellum* is a cellulolytic gram-positive spore-forming bacterium and is able to grow on cellulose as a sole carbon source. It is capable of fermenting a variety of cellulosic substrates, glucose, cellobiose, but not the xylan or pentose sugars, mainly xylose and xylobiose generated from hemicellulose. The fermentation products usually include ethanol, lactate and acetate. The ratio between the products varies based on strains, substrates and cultivation conditions. The cellulase of *C. thermocellum* is strongly inhibited by cellobiose accumulation. Although *C. thermocellum* has the potential for the direct conversion of cellulose to ethanol, the slow growth rate coupled with the limited yield and tolerance to ethanol has limited the development of a practical fermentation process (Jones and Woods, 1989). The use of mixed culture systems is of great interest. *C. thermocellum* has been cocultivated with other thermophilic, anaerobic bacteria that are capable of utilizing pentoses, hexoses, and cellobiose, e.g., *C. thermohydrosulfuricum* (Germain et al., 1986; NG et al., 1981), *C. thermosaccharolyticum* (Saddler and Chan, 1984; Venkateswaren and Demain, 1986; Wang et al., 1983), *Thermoanaerobium brockii* (Lamed and Zeikus, 1980), *Thermoanaerobacter ethanolicus* (Wiegel and Ljungdahl, 1981), and *Bacillus stearothermophilus* (Shama, 1991).

It was found that coculture significantly increased the cellulosic substrate consumption and ethanol yield and rate. For example, Ng et al. (1981) used cocluture of *C. thermocellum* and *C. thermohydrosulfuricum* for fermenting MN300 cellulose and achieved ethanol yield of 90% theoretical yield. They also obtained three-fold increase in the ethanol production rate and two-fold decrease in the acetate production rate.

Table 6. Advantages of using *C. thermocellum* for ethanol fermentation from biomass (Adapted from Demain et al., 2005)

- The cellulolytic and xylanolytic ethanogenic nature, allowing saccharification and fermentation in a single step
- The anaerobic nature, negating the need for expensive oxygen transfer
- Low cell growth yield, favoring ethanol conversion
- The thermophilic nature, facilitating ethanol removal and recovery
- The thermophilic nature, reducing cooling cost
- Thermophilic fermentation being less prone to contamination
- Thermophilic biomass-degrading enzymes enhancing protein stability
- Amenability to coculture with other ethanol-producing and pentose fermenting organisms

Table 7. Ethanol producing microorganisms and companies that have plan to use them in their processes (Adapted from Drapcho et al., 2008)

Microorganism	Company
Z. mobilis	DuPont, Poet (formely Broin)
S. cerevisiae	Iogen
S. cerevisiae	Nedalco, Mascoma, SunOpta
E. coli	Verenium Corp. (formerly Celunol, BC International), BioEthanol Japan
K. oxytoca	Verenium Corp.
B. stearothermophilus	Colusa Biomass
T. mathranii	Biogasol
C. phytofermentans	SunEthanol

In addition to bacteria, thermotolerant and thermophilic yeast strains were also studied for the ethanol production in SSF, e.g. *Fabospora fragilis*, *Saccharomyces uvarum*, *Saccharomyces carlsbergensis*, *Kluyveromyce fragilis*, *Candida brassicae*, *Candida lusitaniae*, and *Kluyveromyces marxianus* have been evaluated for future use in SSF (Ballesteros et al., 1991, 2004; Hong et al., 2007; Krishna et al., 2001; Spindler et al., 1988, 1989b; Szczodrak and Targonski, 1988).

However, studies on the fermentation utilizing thermotolerant and thermophilic organism have shown this process to be very slow with limited ethanol yield, which most probably is due to the low resistance of organisms to high concentration of ethanol (Lin and Tanaka, 2006). Another problem of this process is the production of various competing by-products, mainly acetate and lactate, particularly in the case of bacterial fermentation.

As shown in Table 7, several ethanologenic microorganisms have been developed and will be employed by some companies to produce cellulosic ethanol.

In sum, all studied ethanol producing microbial system (i.e., microorganisms and compatible combinations of microorganisms) fall short of the ideal traits suitable for industrial cellulosic ethanol production as described above. Most commonly, this shortfall is due to one or more of several essential limitations: (i) lack ability to synthesize a cellulase and/or xylanase system at high activities; (ii) an inability to utilize the range of carbohydrates present in biomass (e.g. cellulose, xylan) while also producing ethanol at high yield and high selectivity; (iii) distinct requirements for oxygen for various functions important to the process (Lynd, 1996).

For example, the microorganisms with the first limitation include *E. coli* KO11, *Z. mobilis*, and commercially-used wild *S. cerevisiae*. Example of microorganism with the second limitation is *C. thermocellum*. The typical microorganisms with the third limitation are fungi such as *T. reesei* and *A. niger*. These fungi are unable to grow and produce cellulases under anaerobic conditions so that cellulase production by these fungi can't happen with the ethanol fermentation by ethanologenic microorganisms in the same bioreactors. Addressing the above limitations is a key focus for improvement and development of microorganisms for successful SSF, SSCF and/or CBP processes in industrial ethanol production.

CONCLUSION

Although significant progress has been made on the bioconversion of lignocellulosic biomass into fuel ethanol during the past years, commercialization of cellulosic ethanol has been difficult, not only due to the recalcitrant nature of biomass itself, but also due to multiple costly treatments required for effective processing. Future research emphasis could be primarily put on three key steps such as biomass pretreatment, enzymatic hydrolysis and fermentation.

Prospective Pretreatment Technologies

The development of cost-effective biomass pretreatment technologies can significantly ease the difficulties of biomass hydrolysis and fermentation and reduce the overall production cost of cellulosic ethanol. Future research could be aimed at: first, reduction of water and chemical use for environmental protection; second, recovery of carbohydrates and by-products for efficiently fractionating biomass into multiple streams containing fermentable sugars and value-added by-products in concentrations that make purification, utilization and/or recovery economically practical are likely to gain research interest; third, development of clean delignification for selective lignin-removal, remaining cellulose and hemicellulose intact for downstream bioconversion processes; fourth, fundamental understanding of pretreatment mechanisms and the relationship between the biomass structure features and enzymatic hydrolysis; fifth, reduction of the generation of inhibitors for fermentation and/or enzymatic hydrolysis to reduce or eliminate costly detoxification processes.

Advanced Cellulase Enzyme

Reducing the cost of cellulase enzyme production is a critical issue in the enzymatic hydrolysis of lignocellulosic materials. It will finally result in the reduction of the cost of cellulosic ethanol via the reduction of enzymatic hydrolysis contribution to the overall ethanol production cost.

This aggressive but achievable goal can be approached from five aspects in general by making significant improvement in: (1) enzyme productivity (reduced cost per unit enzyme); (2) cellulase activity (reduced dose of enzyme needed); (3) tolerant to wide range of pH and high temperature; (4) tolerant to inhibitors produced during pretreatment; (5) resistant to nonproductive and irreversible adsorption to lignin. Genetic engineering have been used to clone the cellulase coding genes into yeasts, fungi, bacteria, and plants to create new cellulase production systems with possible improvement of enzyme production and activity and reduction of enzyme production cost. In addition, many genes encoding previously unknown *T. reesei* proteins were found to have important effects on the induction of cellulase. These findings and future characterizations of the genes will likely have considerable implications for development of processes for industrial production of cellulases.

Promising Prospects of Ethanol Fermentation

Ethanol fermentation involves several significant challenges, such as the necessity of converting both hexose and pentose sugars into ethanol, the incompatibility between ethanologenic microorganism and hydrolytic enzymes, and the low rates of cellulose hydrolysis. There are still challenges needed to be further addressed even though great achievements have been recently made on the development of improved ethanol fermentation processes. (1) Cofermentation of hexose and pentose sugars into ethanol with high yield; (2) Improvement of tolerance of microorganisms to inhibitors derived from pretreatment processes; (3) Compatibility (pH and temperature) between enzymes and microorganism within SSF processes; (4) Cellulolytic and xylanolytic ethanol producing microorganisms for realization of CBP processes; (5) immobilization of fermenting microorganisms for eliminating removal and recycle of microorganisms from downstream product and benefiting the reuse of microorganisms. It could also significantly reduce the loss of intracellular enzyme activity and increase the endurable sugar concentration level of fermenting microorganisms (Najafpour, 1990; Yamada et al., 2002); (6) reduction of the need of sterilization for fermentation processes.

REFERENCES

Ahring, BK; Nielsen, K; Bjerre, AB; Schmiidt, AS. *Bioresour. Technol.*, 1996, 58, 107-113.

Ander, P; Eriksson, KE. *In Progress in Ind. Microbio*l; Bull, MJ; Ed; Elsevier: New York, NY, 1978; Vol. 14, pp 1.

Aristidou, A; Penttila, M. *Curr. Opin. Biotechnol.*, 2000, 11, 187-198.

Baker, JO; Adney, WS; Nieves, RA. *Appl. Biochem. Biotechnol.*, 1994, 45/46, 245-256.

Ballesteros, I; Ballesteros, M; Cabanas, A; Carrasco, J; Martin, C; Negro, MJ; Saez, F; Saez, R. *Appl. Biochem. Biotechnol.*, 1991, 28/29, 307-316.

Ballesteros, M; Oliva, JM; Manzanares, P; Negro, MJ; Ballesteros, I. *World J. Microbiol. Biotechnol.*, 2002, 18, 559-561.

Ballesteros, M; Oliva, JM; Negro, MJ; Manzanares, P; Ballesteros, I. *Process Biochem.*, 2004, 39, 1843-1848.

Beardmore, DH; Fan, LT; Lee, YH. *Biotechnol. Lett.*, 1980, 2, 435-438.

Beldman, G; Rombouts, FM; Voragen, AGJ., Pilnik, W. *Enzyme Microb. Technol.*, 1984, 6, 503-507.

Bothast, RJ; Saha, BC; Flosenzier, AV; Ingram, LO. *Biotechnol. Lett.*, 1994, 16, 401-406.

Bothast, RJ; Nichols, NN; Dien, BS; Cotta, MA. *Society of Industrial Microbiology Annual Meeting*, 2002, paper No. 12-1.

Boussaid, A; Robinson, J; Cai, YJ; Gregg, DJ; Saddler, JR. *Biotechnol. Bioeng.*, 1999, 64, 284-289.

Brat, D; Boles, E; Wiedemann, B. *Appl. Environ. Microbiol.*, 2009, 75, 2304-2311.

Breuil, C; Chan, M; Saddler, JN. *Appl. Microbiol. Biotechnol.*, 1990, 34, 31-35.

Breuil, C; Chan, M; Gilbert, M; Saddler, JN. *Bioresour. Technol.*, 1992, 39, 139-142.

Brooks, TA; Ingram, LO. *Biotechnol. Progr.*, 1995, 11, 619-625.

Browner, C. (2000). Remarks as prepared for delivery to press conference on March 20, 2000. http://www.epa.gov/otaq/consumer/fuels/mtbe/ press34b.pdf.

Campbell, CJ; Laherrere, JH. *Sci. Am.*, 1998, 3, 78-83.

Castanon, M; Wilke, CR. *Biotechnol. Bioeng.*, 1981, 23, 1365-1372.

Chammem, N; Kachouri, M; Mejri, M; Peres, C; Boudabous, A; Hamdi, M. *Bioresour. Technol.*, 2005, 96, 1311-1316.

Chandra, RP; Bura, R; Mabee, WE; Berlin, A; Pan, X; Saddler, JN. *Adv. Biochem. Engin/Biotechnol.*, 2007, 108, 67-93.

Chang, V; Holtzapple, M. *Appl. Biochem. Biotechnol.*, 2000, 84/86, 5-37.

Cheung, SW; Anderson, BC. *Bioresour. Technol.*, 1997, 59, 81-96.

Clark, TA; Mackie, KL. J. *Wood Chem. Technol.*, 1987, 7, 373-403.

Converse, AO; Matsuno, R; Tanaka, M; Taniguchi, M. *Biotechnol. Bioeng.*, 1988, 32, 38-45.

Converse, AO; Optekar, JD. *Biotechnol. Bioeng.*, 1993, 42, 145-148.

Corredor, DY; Sun, XS; Salazar, JM; Hohn, KL; Wang, DJ. *Biobased Mater. Bioenergy*, 2008, 2, 43-50.

Coughlan, MP; Ljungdahl, LG. In Biochemistry and Genetics of Cellulose Degradation; Aubert, JP; Beguin, P; Millet, J; Eds; Academic Press Inc.: Orlando, FL, 1988, 11-30.

Deanda, K; Zhang, M; Eddy, C; Picataggio, S. *Appl. Environ. Microbial.*, 1996, 62, 4465-4470.

De Bari, I; Nanna, F; Braccio, G. *Ind. Eng. Chem. Res.*, 2007, 46, 7711-7720.

Dekker, RF; Karageorge, H; Wallis, AF. *A. Biocatalysis*, 1987, 1, 45-59.

Demain, AL; Newcomb, M; Wu, JHD. *Microbiol. Mol. Biol. Rev.*, 2005, 69, 124-154.

Dewes, T; Hünsche, E. *Biol. Agric. Hortic.*, 1998, 16, 251-268.

Dien, BS; Hespell, RB; Wyckoff, HA; Bothast, R. J. *Enzyme Microb. Technol.*, 1998, 23, 366-371.

Dien, BS; Nichols, NN; Obryan, PJ; Bothast, RJ. *Appl. Biochem. Biotechnol.*, 2000, 84, 181-196.

Dien, BS; Cotta, MA; Jeffries, TW. *Appl. Microbiol. Biotechnol.*, 2003, 63, 258-266.

Din, N; Damude, HG; Gilkes, NR; Miller, RC; Warren, RAJ; Kilburn, DG. *Proc. Natl. Acad. Sci.*, 1994, 91, 11383-11387.

Ding, H; Vlasenko, E; Shoemaker, CF; Shoemaker, S. In Food for Health in the Pacific Rim; Whitaker, JR; Haard, NF; Shoemaker, CF; Singh, RP; Eds; Wiley-Blackwell: *Trumbull, CT*, 1998, 433-444.

Doran, JB; Aldrich, HC; Ingram, LO. *Biotechnol. Bioeng.*, 1994, 44, 240-247.

Doran, JB; Caripe, J; Sutton, M; Foster, B. *Appl. Biochem. Biotechnol.*, 2000, 84-86, 141-152.

Drapcho, CM; Nhuan, NP; Walker, TH. Ethanol production. *Biofuels Engineering Process Technology; McGraw-Hill*, New York, NY, 2008, 105-195.

Duff, SJB; Murray, WD. *Bioresour. Technol.*, 1996, 55, 1-33.

Eklund, R. Ph.D. thesis, Chemical Engineering 1, *Lund Institute of Technology*, Lund, Sweden, 1994.

Energy information administration (EIA). http://www.eia.doe.gov/basics/ quickoil.html. 2009.

Eriksson, T; Borjesson, J; Tjerneld, F. *Enzyme Microb. Technol.*, 2002, 31, 353-364.

Excoffier, G; Toussaint, B; Vignon, MR. *Biotechnol. Bioeng.*, 1991, **38**, 1308-1317.

Fan, LT; Lee, Y; Gharpuray, MM. *Adv. Biochem. Eng.*, 1982, 23, 157-187.

Fan, LT; Gharpuray, MM; Lee, YH. Cellulose Hydrolysis Biotechnology Monographs. *Springer: Berlin*, 1987, 57.

FAO Corporate Document Repository. (2009). http://www.fao.org/ docrep/w7241e/w 7241e08.htm.

Fein, JE; Potts, D; Good, D; Beavan, M; O'Boyle, A; Dahlgren, D; Beck, MJ; Griffith, RL. *Energy Biomass and Waste*., 1991, 15, 745-765.

Fogler, HS. Elements of chemical reaction engineering. Prentice Hall. Englewood Cliffs, N. J., 1986.

Freer, SN; Detroy, RW. *Biotechnol. Bioeng*., 1983, 25, 541-557.

Fujita, Y; Ito, J; Ueda, M; Fukuda, H; Kondo, K. *Appl. Environ. Microbiol*., 2004, 70, 1207-1212.

Fukaya, Y; Hayashi, K; Wada, M; Ohno, H. *Green Chem*., 2008, 10, 44-46.

Genomics: GTL. (2009). http://genomicsgtl.energy.gov/biofuels/ transportation.shtml.

Germain, P; Toukourou, F; Donaduzzi, L. *Appl. Microbiol. Biotechnol*., 1986, 24, 300-305.

Ghose, TK; Ghosh, PJ. *Appl. Chem. Biotechnol*., 1978, 28, 309-320.

Ghose, TK; Bisaria, VS. *Biotechnol. Bioeng*., 1979, 21, 131-146.

Girard, DJ; Converse, AO. *Appl. Biochem. Biotechnol*., 1993, 39/40: 521-533.

Gottschalk, G. Bacterial metabolism; 2nd edn. Springer: New York, NY, 1986, 237.

Gregg, DJ; Saddler, JN. *Biotechnol. Bioeng*., 1996, 51, 375-383.

Gulati, M; Westgate, PJ; Brewer, M; Hendrickson, R; Ladisch, MR. *Appl. Biochem. Biotechnol*., 1996, 57/58, 103-119.

Gunasekaran, P; Raj, KC. Ethanol fermentation technology - *Zymomonas mobilis*. http:// www.ias.ac.in/currsci/jul10/articles14.htm. 1999.

Gusakov, AV; Sinitsyn, AP; Manenkova, JA; Protas, OV. *Appl. Biochem. Biotechnol*., 1992, 34/35, 625-637.

Hartely, BS; Shama, G; Philos. Trans. R. *Soc. Lond. A*., 1987, 321, 555-568.

Hartely, BS. Thermophilic ethanol production in a two-stage closed system. Patent U.S. 5.182,299. 1993.

Helle, SS; Duff, SJB; Cooper, DG. *Biotechnol. Bioeng*., 1993, 42, 611-617.

Herrera, A; Tellez-Luis, SJ; Ramirez, JA; Vazquez, M. *J Cereal Sci*., 2003, 37, 267-274.

Hinman, ND; Schell, DJ; Riley, CJ; Bergeron, PW; Walter, P. *J. Appl. Biochem. Biotechnol*., 1992, 34/35, 639-649.

Ho, NWY; Chen, Z; Brainard, A. *Appl. Environ. Microbiol*., 1998, 64, 1852-1859.

Ho, NWY; Chen, Z; Brainard, AP; Sedlak, M. Adv. *Biochem. Eng./Biotechnol*., 1999, 65, 163-192.

Holtzapple, MT; Cognata, M; Shu, Y; Hendrickson, C. *Biotechnol. Bioeng*., 1990, 36, 275-287.

Hong, J; Wang, Y; Kumagai, H; Tamaki, H. *J. Biotechnol*., 2007, 130, 114-123.

Horton, GL; Rivers, DB; Emert, GH. Ind. Eng. Chern. *Prod. Res. Dev*., 1980, 79, 422-429.

Hsu, TA; Ladisch, MR; Tsao, GT. *Chem. Technol*., 1980, 10, 315-319.

Hsu, TA. In Handbook on Bioethanol, Production and Utilization; Wyman, CE; Ed; Taylor and Francis: Washington, DC, 1996, 179-212.

Huang, XL; Penner, MH. *J. Agric. Food Chem*., 1991, 39, 2096-2100.

Hwang, SS; Lee, SJ; Kim, HK; Ka, JO; Kim, KJ; Song, HG. *J. Microbiol. Biotechnol*., 2008, 18, 1819-1825.

Ingram, LO; Conway, T; Clark, DP; Sewell, GW; Preston, J. *F. Appl. Environ. Microbiol.*, 1987, 53, 2420-2425.

Ingram, LO; Doran, J. *FEMS Microbiol. Rev.*, 1995, 16, 235-241.

Ingram, LO; Gomez, PF; Lai, X; Moniruzzaman, M; Wood, BE; Yomano, LP; York, SW. *Biotechnol. Bioeng.*, 1998, 58, 204-214.

Ingram, LO; Aldrich, HC; Borges, ACC; Causey, TB; Martinez, A; Morales, F; Saleh, A; Underwood, SA; Yomano, LP; York, SW; Zaldivar, J; Zhou, SD. *Biotechnol. Progr*, 1999, 15, 855-866.

Ishihara, M; Uemura, S; Hayashi, N; Shimizu, K. *Biotechnol. Bioeng.*, 1991, 37, 948-954.

Iyer, PV; Wu, ZW; Kim, SB; Lee, YY. *Appl.* Biochem. Biotechnol., 1996, 57/58, 121-132.

Jarboe, LR; Grabar, TB; Yomano, LP; Shanmugan, KT; Ingram, LO. *Adv. Biochem. Engin/Biotechnol.*, 2007, 108, 237-261.

Jeffries, TW; Shi, N. Adv. *Biochem. Eng./Biotechnol.*, 1999, 65, 118-161.

Jeffries, TW. *Nat Biotechnol.*, 2005, 23, 40-41.

Jones, DT; Woods, DR. In Clostridia. Minto, N. P; Clarke, D. J; Eds; *Plenum Publishing Corporation*, New York, NY, 1989, Vol. 3, 105-144.

Kadam, KL; Camobreco, VJ; Glazebrook, BE; Forrest, LH; Jacobson, WA; Simeroth, DC; Blackburn, WJ; Nehoda, KC. Environmental Life Cycle Implications of Fuel Oxygenates Production from California Biomass, NREL/TP-580-25688, *National Renewable Energy Laboratory: Golden, CO.* 1999.

Kadam, KL; Rydholm, EC; Mcmillan, JD. *Biotechnol. Prog.*, 2004, 20, 698-705.

Katahira, S; Fujita, Y; Mizuike, A; Fukuda, H; Kondo, A. *Appl. Environ. Microbiol.*, 2004, 70, 5407-5414.

Kaya, F; Heitmann, JA; Joyce, TW. *Tappi J.*, 1995, 78, 150-157.

Kim, S; Holtzapple, MT. *Bioresour. Technol.*, 2005, 96, 1994-2006.

Kim, S; Holtzapple, MT. *Bioresour. Technol.*, 2006, 97, 583-591.

Kim, TH; Kim, JS; Sunwoo, C; Lee, YY. *Bioresour. Technol.*, 2003, 90, 39-47.

Knauf, M; Moniruzzaman, M. INT. *Sugar J.*, 2004,106, 147-150.

Kohlmann, KL; Sarikaya, A; Westgate, PJ; Weil, J; Velayudhan, A; Hendrickson, R; Ladisch, MR. In Enzymatic Degradation of Insoluble Carbohydrates; Saddler, JN; Penner, MH; Eds; Springer: *Netherlands*, 1995, 237-255.

Kosan, B; Michels, C; Meister, F. *Cellulose*, 2008, 15, 59-66.

Krishna, SH; Reddy, TJ; Chowdary, GV. *Bioresour, Technol.*, 2001, 77,193-196.

Kristensen, JB; Thygesen, LG; Felby, C; Jørgensen, H; Elder, T. *Biotechnol. Biofuels.*, 2008, 1, 1-5.

Kuyper, M; Winkler, AA; van Dijken, JP; Pronk, J. *T. FEMS Yeast Res.*, 2004, 4, 655-664.

Kuyper, M; Hartog, MMP; Toirkens, MJ; Almering, MJH; Winkler, AA; van Dijken, JP; Pronk, J. *T. FEMS Yeast Res.*, 2005, 5, 399-409.

Ladisch, MR; Voloch, M; Hong, J; Bienkowski, P; Tsao, GT. IandEC *Process Des. Dev.*, 1984, 23, 437-443.

Ladisch, MR; Dyck, K. *Science*, 1979, 205, 898-900.

Lamed, R; Zeikus, JG. *J. Bacteriol.*, 1980, 144, 569-578.

Larsen, L; Nielsen, P; Ahring, BK. *Arch. Microbiol.* 1997, 168, 114-119.

Lee, J. *Journal of Biotechnology*, 1997, 56, 1-24.

Lee, JW; Gwak, KS; Park, JY; Park, MJ; Choi, DH; Kwon, M; Choi, IG. *J. Microbiol.*, 2007, 45, 485-491.

Li, A; Antizar-Ladislao, B; Khraisheh, M. *Bioprocess. Biosyst. Eng.*, 2007, 30, 189-196.
Lin, Y; Tanaka, S. Appl. *Microbiol. Biotechnol.*, 2006, 69, 627-642.
Lynd, LR; Grethlein, HE; Wolkin, RH. *Appl. Environ. Microbiol.*, 1989, 55, 3131-3139.
Lynd, LR. Annu. Rev. *Energy Environ.*, 1996, 21, 403-465.
Lynd, LR; Weimer, PJ; van Zyl, WH; Pretorius, IS. *Microbiol. Mol. Biol. Rev.*, 2002, 66, 506-577.
Lynd, LR; Van Zyl, WH; McBride, JE; Laser, M. *Curr. Opin. Biotechnol.*, 2005, 16, 577-583.
Mandels, M; Hontz, L; Nystrom, J. *Biotechnol. Bioeng.*, 1974, 16, 1471-1493.
Mandels, M; Medeiros, JE; Anderotti, RE; Bissett, FH. *Biotehnol. Bioeng.*, 1981, vol. XXIII, 2009-2026.
Martin, RS; Bushell, D; Leak, DJ; Hartley, BS. *Biotechnol. Bioeng.*, 1993, 44, 21-28.
McMillan, JD. In Enzymatic Conversion of Biomass for Fuels Production, Himmel, ME; Baker, JO; Overend, RP; Eds; American Chemical Society: Washington, DC, 1994, 292-324.
Medve, J; Karlsson, J; Lee, D; Tjerneld, F. *Biotechnol. Bioeng.*, 1998, 59, 621-634.
Mes-Hartree, M; Dale, BE; Craig, WK. *Appl. Microbiol. Biotechnol.*, 1988, 29, 462-468.
Mes-Hartree, M; Hogan, CM; Saddler, JN. *Biotechnol. Bioeng.*, 1987, 30, 558-564.
Mohagheghi, A; Evans, K; Finkelstein, M; Zhang, M. *Appl. Biochem. Biotechnol.*, 1998, 70/72, 285-299.
Moniruzzaman, M; Dien, BS; Ferrer, B; Hespell, RB; Dale, BE; Ingram, LO; Bothast, R. *J. Biotechnol. Lett.*, 1996, 18, 955-990.
Mononmani, HK; Sreekantiah, KR. *Enzyme Microb. Technol.*, 1987, 9, 484-488.
Mosier, N; Wyman, C; Dale, B; Elander, R; Lee, YY; Holtzapple, M; Ladisch, M. *Bioresour. Technol.*, 2005, 96, 673-686.
Murnen, HK; Balan, V; Chundawat, SPS; Bals, B; Sousa, LC; Dale, BE. *Biotechnol. Prog.*, 2007, 23, 846-850.
Najafpour, GD. *J Sci Islam Repub Iran*, 1990, 1, 172-176.
Ng, TK; Ben-Bassat, A; Zeikus, JG. *Appl. Environ. Microbiol.* 1981, 41, 1337-1343.
Nghiem, NP; Davison, BH. Continuous ethanol production using immobilized-cell/enzyme biocatalysts in fluidized-bed reactor (FBR). Oak Ridge National Laboratory report number, 2002. ORNL/TL-2002/164.
Nieves, RA; Ehrman, CI; Adney, WS; Elander, RT; Himmel, ME. *World, J. Microbiol. Biotechnol.*, 1998, 14, 301-304.
Ohta, K; Beall, DS; Mejia, JP; Shanmugam, KT; Ingram, LO. *Appl. Environ. Microbiol.*, 1991a, 57, 893-900.
Ohta, K; Beall, DS; Mejia, JP; Shanmugam, KT; Ingram, LO. *Applied and Environmental Microbiology*, 1991b, 57, 2810-2815.
Onuki, S; Koziel, KA; van Leeuwen, J; Jenks, WS; Grewell, D; Cai, LS. ASABE, paper No. 085136, Providence, RI, 2008.
Ooshima, H; Sakata, M; Harano, Y. *Biotechnol. Bioeng.*, 1986, XXVIII, 1727-1734.
Op Den Camp, MJH; Harhangi, RH; Van Der Drift, C. Patent, WO 03/062430 A1. 2003.
Palonen, H; Thomsen, AB; Tenkanen, M; Schmidt, AS; Viikari, U. *Appl. Biochem. Biotechnol.*, 2004, 117, 1-17.
Pan, X; Xie, D; Kang, KY; Yoon, SL; Saddler, JN. *Appl. Biochem. Biotechnol.*, 2007, 136/140, 367-377.

Pan, X; Xie, D; Yu, YW; Saddler, JN. *Biotechnol. Bioeng*., 2008, 101, 39-48.
Penner, MH; Liaw, ET. In Enzymatic Conversion of Biomass for Fuels Production; Himmel, M. E; Baker, J. O; Overend, R.P; Eds; American Chemical Society: *Washington, DC*, 1994, 363-371.
Philippidis, GP; Spindler, DD; Wyman, CE. *Appl. Biochem. Biotechnol*., 1992, 34-35, 543-556.
Philippidis, GP; Smith, TK; Wyman, CE. *Biotechnol. Bioeng*., 1993, 41, 846-853.
Prior, BA; Day, DF. *Appl. Biochem. Biotechnol*., 2008, 146, 151-164.
Ramos, JP; Breuil, C; Saddler, JN. *Enzyme Microb. Technol*., 1993, 15, 19-25.
Ramos, LP; Saddler, JN. *Appl. Biochem. Biotechnol*., 1994, 45/46, 193-207.
Rani, KS; Swamy, MV; Seenayya, G. *Process Biochem*., 1998, 33, 435-440.
Rapp, P; Beerman., A. In Biosynthesis and biodegradation of cellulose; Haigler, CH; Weimer, PJ; Eds; Marcel Dekker, Inc: New York, NY, 1991; pp 535-595.
Reese, ET; Sui, RGH; Levinson, HSJ. *Bacteriol*., 1950, 59, 485-497.
Reese, ET. *Appl. Microbiol*., 1956, 4, 39-45.
Reshamwala, S; Shawky, BT; Dale, BE. *Appl. Biochem. Biotechnol*., 1995, 51-52: 43-55.
Rogalinski, T; Ingram, T; Brunner, G. J. *Supercrit. Fluids*, 2008, 47, 54-63.
Rogers, PL. K; Lee, J; Skotnicki, ML; Tribe, DE. *Advances Biochem. Eng*., 1982, 23, 37-84.
Ropare, M; Marchal, R; Pourquie, J; Vandecasteele, J. P. *Bioresour. Technol*., 1992, 42, 197-204.
Rudolf, A; Alkasrawi, M; Zacchi, G; Liden, G. Enzyme Microb. *Technol*., 2005, 37, 195-204.
Runquist, D; Fonseca, C; Radstrom, P; Spencer-Martins, I; Hahn-Hagerdal, B. *Appl. Microbiol. Biotechnol*., 2009, 82, 123-130.
Ruohonen, L; Aristidou, A; Frey, AD; Penttila, M; Kallio, PT. Enzyme Microb. *Technol*., 2006, 39, 6-14.
Ruth, M.; Lbsen, K., 2004. Bioethanol Production using Enzymatic Processes: 20 Year Economic Outlook. *http://www.p2pays.org/ref/40/39143.pdf*. Accessed in October, 2009.
Rydholm, SA. Pulping Processes; Interscience Publishers, John Wiley and Sons: New York, NY, 1965, 3-90.
Saddler, JN; Brownell, HH; Clermont, LP; Levitin, N. *Biotechnol. Bioeng*., 1982, 24, 1389-1402.
Saddler, JN; Chan, MKH. Can. *J. Microbiol*., 1984, 30, 212-220.
Saddler, JN; Gregg, DJ. In Forest Products Biotechnology; Bruce, A; Palfreyman, JW; Eds; Taylor and Francis Ltd: London, 1998, 183-207.
Saha, BC. In Lignocellulose Biodegradation; American Chemical Society: *Washington, DC*, 2004, 2-34.
Saha, BC; Biswas, A; Cotta, MAJ. *Biobased Mater. Bioenergy*, 2008, 3, 210-217.
Schell, DJ; Farmer, J; Newman, M; McMillan, JD. *Appl. Biochem. Biotechnol*., 2003, 105/108, 69-85.
Schurz, J. In Bioconversion of Cellulosic Substances into Energy Chemicals and Microbial Protein Symp. Proc; Ghose, TK; Ed; New Delhi, IIT: Delhi, 1978; pp 37.
Schwarz, WH. *Appl. Microbiol. Biotechnol*., 2001, 56, 634-649.
Sedlak, M; Ho, N. W. Y. *Appl. Biochem. Biotechnol*., 2004, 113/116, 403-416.
Shama, G. *Biotechnol. Lett*., 1991, 13, 761-764.
Shi, J; Chinn, MS; Sharma-Shivappa, RR. *Bioresour. Technol*., 2008, 99, 6556-6564.
Shin, HS; Zeikus, JG; Jain, MK. Appl. *Microbiol. Biotechnol*., 2002, 58, 476-481.

Silverstein, RA; Chen, Y; Sharma-Shivappa, RR; Boyette, MD; Osborne, J. *Bioresour. Technol.*, 2007, 98, 3000-3011.
Singh, A; Kumar, PKR; Schügerl, K. *J. Biotechnol.*, 1991, 18, 205-212.
Singh, P; Suman, A; Tiwari, P; Arya, N; Gaur, A; Shrivastava, AK. World *J. Microbiol. Biotechnol.*, 2008, 24, 667-673.
Spano, L; Allen, A; Tassinari, T; Mandels, M; Ryu, DDY. In Proceeding of the Second Annual Symposium on Fuels from Biomass; Shuster, W.W; Ed; *John Willey and Sons*: New York, NY, 1978, 671-684.
Spindler, DD; Wyman, CE; Mohagheghi, A; Grohmann, K. *Appl. Biochem. Biotechnol.*, 1988, 17, 279-294.
Spindler, DD; Wyman, CE; Grohmann, K; Mohagheghi, A. *Appl. Biochem. Biotechnol.*, 1989a, 20-21, 529-540.
Spindler, DD; Wyman, CE; Grohmann, K. *Biotehnol. Bioeng.*, 1989b, 34, 189-195.
Spindler, DD; Wyman, CE; Grohmann, K; Philippidis, P. *Biotechnol. Lett.*, 1992, 14, 403-407.
Sprenger, GA. FEMS Microbiol. *Lett.*, 1996, 145, 301-307.
Sternberg, D. Biotechnol. *Bioeng. Symp.*, 1976, 6, 35-53.
Sues, A; Millati, R; Edebo, L; Taherzadeh, M. *J. FEMS Yeast Res.*, 2005, 5, 669-676.
Sun, Y; Cheng, JY. *Bioresour. Technol.*, 2002, 83, 1-11.
Szczodrak, J; Targonski, Z. *Biotechnol. Bioeng.*, 1988, 31, 300-303.
Szczodrak, J; Targonski, Z. *Acta Biotechnol.*, 1989, 6, 555- 564.
Taherzadeh, MJ; Karimi, K. Int. *J. Mol. Sci.*, 2008, 9, 1621-1651.
Tan, LUL; Yu, EKC; Mayers, P; Saddler, JN. *Appl. Microbiol. Biotechnol.*, 1986, 25, 256-261.
Tan, LUL; Yu, EKC; Mayers, P; Saddler, JN. *Appl. Microbiol. Biotechnol.*, 1987, 26, 21-27.
Teeri, TT. *Trends Biotechnol.*, 1997, 15,160-167.
Teixeira, LC; Linden, JC; Schroeder, HA. *Appl. Biochem. Biotechnol.*, 2000, 84/86: 111-127.
Tengborg, C; Galbe, M; Zacchi, G. *Biotechnol. Prog.*, 2001, 17, 110-117.
Thomas, KC; Hynes, SH; Ingledew, WM. *Process Biochem.*, 1996, 31, 321-331.
Tolan, JS. In Biorefineries-Industrial Processes and Products: Status Quo and Future Directions; Kamm, B; Gruber, P. R; Kamm, M; Eds; WILEY-VCH Verlag GmbH and Co: Hoboken, NJ, 2006, Vol. 1, 193-208.
Tomme, P; Heriban, V; Claeyssens, M. *Biotechnol. Lett.*, 1990, 12, 525-530.
Tshiteya, RM. Conversion technologies-biomass to ethanol-alcohol fuels. National Renewable Energy Lab (NREL), Reference Report No. 3, September. 1992.
Tu, M; Chandra, RP; Saddler, J. *N. Biotechnol. Prog.*, 2007, *23*, 1130-1137.
Väljamäe, P; Sild, V; Nutt, A; Pettersson, G; Johansson, G. Eur. *J. Biochem.*, 1999, 266, 327-334.
Vallander, L; Eriksson, KE. *Enzyme Microb. Technol.*, 1987, 9, 714-720.
Van Wyk, JPH. *Trends Biotechnol.*, 2001, 19, 172-177.
Vazquez, M; Oliva, M; Tellez-Luis, SJ; Ramirez, JA. *Bioresour. Technol.*, 2007, 98, 3053-3060.
Venkateswaren, S; Demain, AL. *Chem. Eng. Commun.*, 1986, 45, 53-60.
Viljoen, JA; Fred, EB; Peterson, WH. *J. Agric. Sci.*, 1926, 16, 1-5.
Wang, DIC; Avgerinos, GC; Biocic, I; Wang, SD; Fang, HY. *Phil. Trans, R. Soc.*, London, 1983, 300, 323-333.

Wang, M; Saricks, C; Santini, D. (1999). Effects of Fuel Ethanol use on Fuel-Cycle Energy and Greenhouse Gas Emissions. Argonne National Laboratory, Argonne, IL.

Warnick, TA; Methé, BA; Leschine; SB. *Int. J. Syst. Evol. Microbiol.*, 2002, 52, 1155-1160.

Wenzel, HF. *The Chemical Technology of Wood.* Academic Press, 1970; pp 32-245.

Wiegel, J; Ljungdahl, LG. *Arch. Microbiol.*, 1981,128, 343-348.

Wiselogel, A; Tyson, S; Johnson, D. In *Handbook on Bioethanol: Production and Utilization*; Wyman, CE; Ed; Taylor and Francis: Washington, DC, 1996, pp 105-118.

Wisselink, HW; Toirkens, MJ; Berriel, MDF; Winkler, AA; van Dijken, JP; Pronk, JT; van Maris, AJ. *A. Appl. Environ. Microbiol.*, 2007, 73, 4881-4891.

Wisselink, HW; Toirkens, MJ; Wu, Q; Pronk, JT; van Maris, AJ. *A. Appl. Environ. Microbiol.*, 2009, 75, 907-914.

Wood, BE; Ingram, LO. *Appl. Environ. Microbiol.*, 1992, 58, 2103-2110.

Wooley, R; Ruth, M; Glassner, D; Sheehan, J. *Biotechnol. Progr.*, 1999, 15, 794-803.

Wright, JD; Power, AJ; Douglas, L. *J. Biotechnol. Bioeng.*, 1986, 17, 285-302.

Wright, JD; Wyman, CE; Grohmann, K. *Appl. Biochem. Biotechnol.*, 1987, 18, 75-90.

Wu, Z; Lee, YY. *Biotech. Lett.*, 1997, 19, 977-979.

Wyman CE; Spindler, DD; Grohmann, K; Lastick, SM. *Biotechnol. Bioeng. Symp.*, 1986, 17, 221-238.

Wyman, CE. *Bioresour. Technol.*, 1995, 50, 3-15.

Wyman, CE. In Handbook on Bioethanol, Production and Utilization; Wyman, C. E; Ed; Taylor and Francis: Washington, DC, 1996, 1-18.

Wyman, CE. Annu. Rev. *Energy Environ.*, 1999, 24, 189-226.

Wyman, CE. *Biotechnol. Prog.*, 2003, 19, 254-262.

Wyman, CE. *MRS Bull.*, 2008, 33, 381-383.

Xin, Z; Qu, Y; Gao, P. *Enzyme Microb. Technol.*, 1993, 15, 62-65.

Yamada, T; Fatigati, MA; Zhang, M. *Appl. Biochem Biotechnol.*, 2002, 98, 899-907.

Yang, B; Wyman, CE. Biotechnol. Bioeng., 2004, 86, 88-98.

Yang, CP; Shen, ZQ; Yu, GC; Wang, J. *Bioresour. Technol.*, 2008, 99, 6240-6245.

Yomano, LP; York, SW; Ingram, LO. *J. Ind. Microbiol. Biotechnol.*, 1998, 20, 132-138.

Yomano, LP; York, SW; Zhou, S; Shanmugam, KT; Ingram, LO. Biotechnol. *Lett.*, 2008, 30, 2097-2103.

Yoon, HH; Wu, Z; Lee, YY. Presented at the 16th Symposium on Biotechnology for Fuels and Chemicals, Gatlinburg, TN. 1994.

Yoshida, M; Liu, Y; Uchida, S; Kawarada, K; Ukagani, Y; Ichinose, H; Kaneko, S; Fukuda, K. Biosci. *Biotechnol. Biochem.*, 2008, 72, 805-810.

Youssef, BM; Aziz, NH. *Cytobios*, 1999, 97, 171-183.

Yu, J; Zhang, JB; He, J; Liu, ZD; Yu, ZN. *Bioresour. Technol.*, 2009, 100, 903-908.

Zhang, C; Eddy, K; Deanda, M. *Science*, 1995, 267, 240-243.

Zhang, YHP; Lynd, LR. *Biotechnol. Bioeng.*, 2004, 88, 798-824.

Zheng, Y; Pan, Z; Zhang, RH; Labavitch, J; Wang, D; Teter, SA; Jenkins, BM. 2007. *Appl. Biochem. Biotechnol.*, 2004, 136/140, 423-435.

Zheng, Y. (2007). Kinetic Modeling of Enzymatic Saccharification and Particleboard Characteristics of Saline Biomass. Ph.D. thesis, Biological and Agricultural Engineering Department, University of California, Davis, USA.

Zheng, Y; Pan, Z; Zhang, RH; Wang, D; Jenkins, B. *Appli. Biochem. Biotechnol.*, 2008, 146, 231-248.

Zheng, Y; Pan, Z; Zhang, RH; Wang, D. *Appl. Energy*, 2009a, 86, 2459-2465.

Zheng, Y; Pan, Z; Zhang, RH; Jenkins, B. *Biotechnol. Bioeng.*, 2009b, 102, 1558-1569.

Zhou, S; Davis, FC; Ingram, LO. Appl. *Environ. Microbiol.*, 2001, 67, 6-14.

Zhu, YM; Lee, YY; Elander, RT. *Appl. Biochem. Biotechnol.*, 2004, 117, 103-114.

In: Ethanol: Production, Cellular Mechanisms and Health Impact ISBN: 978-1-62257-297-7
Editors: Roy I. Henry and Billy P. Woods

Chapter 7

The Effects of Acute, Chronic and Withdrawal from Chronic Ethanol on Emotional Learning

Danielle Gulick and Thomas J. Gould
Department of Psychology, Neuroscience Program,
Temple University, Philadelphia, Pennsylvania, US

Abstract

Alcohol is the most commonly used and abused recreational drug, and one effect of ethanol administration, regardless of whether it is acute or chronic, is the disruption of learning and memory. Although recent studies have demonstrated that ethanol does not produce global deficits but, rather, acts on specific substrates to alter neural function, our understanding of the effects of ethanol on cognitive processes remains incomplete. The studies discussed herein offer support for the specificity of the effects of ethanol on learning-related processes and examine how these effects vary with both the task and the phase of ethanol administration examined. Acute ethanol impairs emotional learning as measured by standard contextual and cued fear conditioning, as well as trace fear conditioning and passive avoidance.

However, the effects of acute ethanol on these tasks are influenced by multiple factors, such as genetics and age. Furthermore, as ethanol administration transitions into chronic and withdrawal from chronic ethanol, the pattern of impairments in emotional learning changes.

This suggests that acute, chronic, and withdrawal from chronic ethanol differentially alter behavior and therefore may also differentially alter neuronal function. Thus, the current review compares and contrasts the effects of acute, chronic, and withdrawal from chronic ethanol within fear conditioning and passive avoidance tasks, and across these two models of aversive/emotional learning.

Keywords: Alcohol, Addiction, Learning and Memory, Hippocampus, Genetics

INTRODUCTION

Despite the well-known deficits in cognition and locomotion caused by both acute and chronic administration of ethyl alcohol (ethanol), approximately 17.6 million adult Americans abuse alcohol on a regular basis (National Institute on Alcohol Abuse and Alcoholism (NIAAA), 2004). Ethanol has an extremely low potency compared to other drugs of abuse, yet 10% of the adult male population in the United States drinks heavily enough to be diagnosed as alcoholic (Crabbe, 1997). Furthermore, this is likely an underestimate of the frequency of alcohol abuse; although the deficits associated with ethanol use are easily identifiable, alcoholism is rarely diagnosed, and few alcoholics seek treatment (for review, Pihl and Peterson, 1995). Successful cessation is associated with immediate improvements in cognitive function, yet the current options for treatment are ineffective; alcoholics are equally likely to successfully quit drinking regardless of whether they are in treatment programs or not (Cutler and Fishbain, 2005). In order to develop more effective therapies, it is important to understand the neural and behavioral changes associated with alcohol abuse.

Whereas individuals may initially consume ethanol for beneficial effects such as reductions in anxiety, neural adaptations occur with chronic use leading to changes in the behavioral effects of ethanol, such as the development of tolerance. In addition, changes in cognitive function occur with alcohol consumption. The cognitive impairments associated with acute and chronic alcohol abuse range from short-term memory loss and learning deficits to blackouts and the permanent inability to form new memories (Ryback, 1971; Wilkinson and Poulos, 1987). Furthermore, withdrawal from chronic ethanol also produces cognitive deficits (Borlikova et al., 2006; Markel et al., 1986; Ripley et al., 2003). Understanding the mechanisms that underlie the cognitive impairments associated with acute, chronic, and withdrawal from chronic ethanol may lead to better treatments for the impairing effects of both acute ethanol intoxication and chronic ethanol abuse.

Initially, it was believed that ethanol created a general depression in the central nervous system, producing a nonspecific disruption in cognitive processing (Himwich, 1932; Wallace, 1932). According to this theory, ethanol-induced performance deficits in tasks that rely on learning or memory were merely one aspect of a generalized neural dysfunction. However, more recent work suggests that the effects of ethanol on learning are highly specific. For example, ethanol has minimal impact on recall of learned information; instead, ethanol impairs the acquisition of learning in many paradigms (Birnbaum et al., 1978; Givens, 1995; Gould, 2003; Higgins et al., 1992), thus producing 'blackouts' for experiences during intoxication.

Ethanol-induced learning deficits have been demonstrated in a wide range of tasks, and understanding the effects of ethanol on learning is an essential step in developing treatments that deal with the many symptoms of alcoholism. This review will focus on emotional learning and, within this field, on aversive classical and instrumental conditioning. Learning how to predict and thus avoid aversive events is essential to the survival of all organisms, and aversive learning can be studied by examining the associations that are formed between environmental stimuli and aversive events. Although these associations are experience-dependent, responses to aversive stimuli are innate; both rats and mice that have been raised entirely in laboratory environments are averse to cues signaling the presence of a predator (Blanchard and Blanchard, 1977; Dell'Omo and Alleva, 1994). Aversive learning, which may reflect or model emotional learning, is seen in all species and is rapidly acquired, making it easy to study in a laboratory setting and externally valid in translation to human research. In addition to being acquired in as little as a single trial, aversive learning is also long-lasting, making it ideal for studying the substrates underlying both short-term and long-term memory.

The two tasks discussed herein are examples of Pavlovian (classical) and instrumental (operant) learning tasks, namely fear conditioning and passive avoidance, that are impaired by pre-training or chronic administration of ethanol (Bammer and Chesher, 1982; Gould, 2003; Melis et al., 1996). Thus, fear conditioning and passive avoidance are useful models for identifying the behavioral and neural correlates of the effects of ethanol on learning. Fear conditioning tests both contextual learning, in which contextual features are associated with an unconditioned stimulus (US) foot shock, and cued learning, in which a discrete conditioned stimulus (CS) cue is associated with the US (Fendt and Fanselow, 1999; Logue et al., 1997; Phillips and LeDoux, 1992). In standard contextual and cued fear conditioning, the animal is placed in a novel context and then exposed to a previously neutral auditory CS that co-terminates with the US; this leads to both context-US and CS-US associations. Both short-term and long-term memory can be tested for contextual and cued learning. In another type of fear conditioning, trace fear conditioning, the CS terminates for a period of time before the presentation of the US (McEchron et al., 1998). In order to learn the association between the CS and the US when there is a delay separating them, the subject must maintain a memory trace of the CS in working memory until presentation of the US. Thus, acquisition of trace fear conditioning may model some aspects of working memory (Carter et al., 2003) that are not recruited by standard fear conditioning. The use of fear conditioning allows investigators to look at multiple types of learning and memory, such as contextual learning, cued learning, and working memory.

In passive avoidance, the animal is placed either on a platform or in one of a connected pair of chambers. Upon stepping down from the platform (step-down avoidance) or moving into the second chamber (inhibitory avoidance), the animal is exposed to an inescapable foot shock. Testing consists of returning the animal to the platform or initial chamber and measuring the latency to step down or into the second chamber (Alkana and Parker, 1979; Bammer and Chesher, 1982). Although there are similarities in the stimuli, behaviors, and neural substrates associated with passive avoidance and the various forms of fear conditioning, there are also differences in the neural substrates underlying these tasks (Wilensky et al., 2000). Thus, both passive avoidance and fear conditioning allow examination of the effects of ethanol on aversive learning and, because some neural substrates involved in these tasks differ, this examination may facilitate identification of the substrates altered by ethanol.

Although the anxiolytic and rewarding properties of ethanol are major factors in addiction, ethanol-associated changes in cognitive processes are also a hallmark of alcohol abuse. In the following sections, we will review the brain areas involved in the effects of ethanol on aversive learning. In addition, we will examine how the acute, chronic, and withdrawal effects of ethanol alter aversive learning, and how methodological differences between studies may influence results. Within our discussion of each phase of ethanol administration, we will review the differential effects of ethanol on fear conditioning and passive avoidance and examine how these differences may inform us about the effects of ethanol on learning. Finally, we will discuss the effects of ethanol on specific neurotransmitter systems as a model for the cellular changes that may underlie ethanol-induced learning deficits.

BRAIN AREAS INVOLVED IN THE MODULATION OF LEARNING BY ETHANOL

The brain regions involved in fear conditioning, which include the hippocampus, amygdala and prefrontal cortex, have been well-characterized (for review, Fendt and Fanselow, 1999). Although less is known about the substrates of passive avoidance, research suggests that the amygdala and hippocampus are also essential for this task (Izquierdo et al., 2006). In both tasks, the hippocampus encodes contextual information (Fanselow, 2000; Martinez et al., 2002; Phillips and LeDoux, 1992; Rudy et al., 2002) and the amygdala is the site where fear-related memories are formed (LeDoux, 2003; Nader et al., 2001; Rodrigues et al., 2004). In addition, the prefrontal cortex mediates the attentional aspects of learning and may underlie attention to conditioned and unconditioned stimuli during training in both tasks (for review, Faw, 2003). Finally, aversive learning is associated with an increase in striatal activity that may reflect anticipation for upcoming events; the striatum may increase attention to predictive stimuli (Matsushima et al., 2003; McNally and Westbrook, 2006) and, thus, may signal to animals that the foot-shock is imminent.

Even though passive avoidance and fear conditioning involve many of the same neural areas, the functions of these areas may differ between tasks. Specifically, the amygdala may be differentially involved in fear conditioning compared to passive avoidance; Wilensky and colleagues (2000) have demonstrated that training day inactivation of the basolateral nucleus of the amygdala inhibits acquisition of fear conditioning but not passive avoidance, while post-training inactivation of the basolateral amygdala alters consolidation of passive avoidance without altering fear conditioning. Thus, understanding the different ways in which ethanol acts in these areas to alter learning may aid in understanding the cognitive changes that occur with alcohol abuse.

Acquisition of aversive learning is consistently altered by ethanol treatment (Givens, 1995; Higgins et al., 1992; Ripley et al., 2003) and this effect may be due to the actions of ethanol in the brain areas discussed above. Ethanol disrupts excitatory activity in the hippocampus (Schummers and Browning, 2001) and amygdala (Roberto et al., 2006), and binge drinking-induced changes in these areas have been implicated in ethanol-induced memory deficits (Stephens et al., 2005). These areas are not independent; instead, they have reciprocal connections that provide feedback for modulation of memory processing. Thus, ethanol may alter activity in one of these structures or may alter the communication between them. It has been suggested that changes in signaling through the pathway connecting the amygdala and hippocampus may underlie some of the ethanol-induced deficits in learning. For example, Abe and colleagues (2004) have demonstrated that ethanol potentiates GABAergic inhibition of synaptic transmission from the basolateral nucleus of the amygdala to the dentate gyrus of the hippocampus. Thus, the effects of ethanol on aversive learning

may be due to changes in the hippocampus and the amygdala, as well as signaling from the amygdala to the hippocampus; this dual process could create additive effects of ethanol on hippocampal function.

The ability of ethanol to act on multiple targets within a single system may also be a factor in the potent effects of ethanol on aversive learning. For example, hippocampus-dependent contextual fear conditioning is particularly sensitive to the impairing effects of ethanol, but cued fear conditioning, a hippocampus-independent form of learning, is impaired by higher doses of ethanol (Gould, 2003; Popke et al., 2000). Furthermore, a hippocampus-dependent form of cued conditioning, trace fear conditioning, also shows greater ethanol-induced deficits than standard cued fear conditioning (Weitemier and Ryabinin, 2003). There are two possible explanations for the greater effects of ethanol on hippocampus-dependent learning processes. It may be that hippocampal substrates underlying learning are more sensitive to the effects of ethanol, while similar substrates in other brain areas are affected to a lesser degree. Alternatively, the greater effects of ethanol on contextual and trace fear conditioning may be due to the dependence of these forms of learning on both the hippocampus and amygdala (Buchel et al., 1999; Logue et al., 1997; Phillips and LeDoux, 1992). As discussed, both hippocampal and amygdalar function are altered by ethanol (Stephens et al., 2005), and the additive effects of ethanol in these areas may lead to the greater deficits seen in contextual and trace fear conditioning compared to cued fear conditioning.

Ethanol may also act in the striatum and prefrontal cortex to alter learning (Vertes, 2006). Ethanol alters striatal neurotransmitter release (Heinz et al., 2004), and it may be that ethanol-induced changes in neurotransmitter release alter signaling from the striatum that is required for anticipatory attention during associative conditioning tasks. Additionally, the prefrontal cortex is involved in attention (for review, Knudsen, 2007) and has been shown to be especially vulnerable to the cellular damage associated with chronic ethanol use (Fadda and Rossetti, 1998).

Therefore, although the effects of ethanol on hippocampal and amygdalar function have received a great deal of attention (Koob, 2004; McBride, 2002), ethanol-induced changes in other, interconnected regions may also disrupt learning (Givens and McMahon, 1997; Lyon et al., 1975; Rezvani and Levin, 2003).

Thus, the acute, chronic, and the post-chronic effects of ethanol in any of these brain areas may produce changes in learning and memory. However, because there are differences in the brain regions (Ambrogi Lorenzini et al., 1991; Wilensky et al., 2000) and molecular substrates (Korte, 2001; Tinsley et al., 2004) involved in fear conditioning and passive avoidance, ethanol may differentially alter learning in these tasks.

EFFECTS OF ACUTE ETHANOL ON AVERSIVE LEARNING

In the following sections, we will first discuss the effects of acute ethanol on acquisition, consolidation, and recall of fear conditioning. We will compare these effects across contextual, cued, and trace fear conditioning as well as across low and high doses of ethanol. We will then discuss the effects of acute ethanol on acquisition, consolidation, and recall of passive avoidance. Finally, we will examine factors, such as development and genetics, that influence the effects of acute ethanol on these tasks.

Standard Fear Conditioning

One of the major effects of ethanol on cognition, whether administered acutely or chronically, is the dose-dependent impairment of learning and memory (Table 1). In fear conditioning, acute ethanol administration produced impairments in contextual and cued fear conditioning by decreasing acquisition of learning (Gould and Lommock, 2003; Gulick and Gould, 2007; Gulick and Gould, 2008b; Melia et al., 1996). Because ethanol blocked acquisition of learning during the training session itself, ethanol-induced learning deficits were visible at 4 hours, 24 hours, and 1 week after training (Gould and Lomock, 2003; Gulick and Gould, 2007). This effect persisted even when ethanol was administered on training and testing days, suggesting that ethanol-induced deficits are not state-dependent (Gould and Lommock, 2003). However, a higher dose of acute ethanol was needed to impair fear conditioning if drug administration occured on both training and testing days compared to training day-only administration (Gould, 2003), suggesting that there may be some effects of ethanol on testing day when ethanol has been administered on training day as well.

A number of studies have demonstrated that ethanol has less of an impact on consolidation and recall of learned information compared to acquisition of information (Ripley et al., 2003; Stephens et al., 2001). Research has demonstrated that a dose of ethanol that impairs learning when administered before training does not alter contextual or cued fear conditioning when administered after training (Gulick and Gould, 2007). Furthermore, Gould (2003) has shown that testing day-only administration of acute ethanol did not disrupt fear conditioning.

Table 1. Effects of Ethanol Treatment on Aversive Learning

Treatment	Trace Conditioning	Contextual Conditioning	Cued Conditioning	Passive Avoidance
	High Doses (1.0+ g/kg acute; variable chronic)			
Acute	Impaired Acquisition[r]	Impaired Acquisition[f g h l]	Impaired Acquisition[a, h]	Impaired[a, i, o]
Chronic	???	???	???	Impaired[c, a, j, k]
Short-term Single Withdrawal	???	No Effect[b]	No Effect[b, p]	Impaired Acquisition and/or Retrieval[a, q]
Short-term Multiple Withdrawal	???	No Effect[b, p]	Impaired with 2+ WDs[b, k, p]	No Effect[n]
Long-term Withdrawal	???	Impaired[d]	No Effect[d]	Variable[c, m, q]
	Low Doses (0.25 g/kg acute)			
Acute	???	Enhanced Acquisition[h]	Enhanced Acquisition[h]	???
Chronic	???	???	???	???
Withdrawal	???	???	???	???

Sources:
a Bammer and Chesher, 1982; b Borlikova et al., 2006; c Casamenti et al., 1993; d Celerier et al., 2000;
e Celik et al., 2005; f Gould, 2003; g Gould and Lommock, 2003; h Gulick and Gould, 2007a;
i Holloway, 1976; j Kuziemka-Leska et al., 1999; k Markel et al., 1986; l Melia et al., 1996; m Melis et al., 1996;
n Popovic et al., 2004; o Rezayof et al., 2007; p Ripley et al., 2003; q Snell and Harris, 1979;
r Weitemier and Ryabinin, 2003.

Taken together, these results suggest that during learning, acute ethanol largely disrupts acquisition. Low doses of ethanol have been shown to produce stimulatory effects in the brain (Gessa et al., 1986; Lima-Landman and Albuquerque, 1989) and, thus, could enhance learning. Furthermore, ethanol has biphasic effects on glutamatergic receptors that may underlie learning, enhancing activity at low doses and inhibiting it at high doses (Lima-Landman and Albuquerque, 1989). Therefore, it is not surprising that ethanol also has biphasic effects on learning in multiple conditioning tasks (Gulick and Gould, 2007; Gulick and Gould, 2008a; Hernandez et al., 1985; Hernandez and Powell, 1986; Hernandez and Valentine, 1990). If low doses of ethanol can enhance associative learning, then these doses may contribute to addiction by facilitating the formation of associations between positive drug effects and the environmental cues that signal drug use. Later re-exposure to the context in which ethanol was consumed may trigger the learned associations and initiate drug-seeking behavior. In support of this, studies have demonstrated that animals develop a conditioned preference for contexts in which they have received ethanol (Cunningham and Noble, 1992; Reid et al., 1985).

Trace Fear Conditioning

Another form of fear conditioning disrupted by ethanol is trace fear conditioning. In trace fear conditioning, there is a delay between the termination of the CS and the onset of the US. In order to learn the association between the CS and the US when there is a delay separating them, animals must keep the CS in working memory. Trace cued fear conditioning is more sensitive to the effects of ethanol than standard cued fear conditioning (Weitemier and Ryabinin, 2003) and there are multiple potential explanations for the differential impact of ethanol on standard and trace cued conditioning. Cued learning may be more sensitive to impairment by ethanol in trace conditioning because the association between the tone and shock is weakened by the delay between them or because these tasks rely on different neural substrates. In support of the latter, Wagner and Hunt (2006) have demonstrated that brief exposure to ethanol (over a period of 5 days; far less than the typical chronic ethanol treatment) impaired trace, but not standard, cued fear conditioning in neonatal rats. This suggests that sensitivities to the effects of ethanol on trace and standard cued fear conditioning develop independently of each other and may be due to the effects of ethanol on separate systems. One region that may be involved in the differential sensitivities of these tasks to ethanol is the hippocampus, which is necessary for trace fear conditioning but not standard cued fear conditioning. Furthermore, Knight and colleagues (2006) demonstrated that awareness – an aspect of cognition mediated by the prefrontal cortex – is necessary for trace, but not standard, cued fear conditioning and, therefore, the effects of ethanol in the prefrontal cortex could produce greater disruption of trace fear conditioning. Thus, the recruitment of different brain regions in trace versus standard cued fear conditioning may make trace conditioning more sensitive to the effects of ethanol.

Acute ethanol clearly disrupts contextual fear conditioning (Gould and Lommock, 2003; Gulick and Gould, 2007) but the disruptive effects of ethanol may depend on the paradigm being used. For example, contextual fear conditioning was less impaired by acute ethanol administration during trace versus standard conditioning (Weitemier and Ryabinin, 2003). The cause underlying the differential effects of ethanol on contextual learning in standard and trace fear conditioning is unclear. These differences may be due to the actions of ethanol in different brain areas or to distinct effects of ethanol in the hippocampus. In support of the latter possibility, there is evidence that trace and standard fear conditioning recruit different hippocampal substrates. In standard contextual fear conditioning, the hippocampus is necessary to learn the context-shock association (Phillips and LeDoux, 1992). In trace fear conditioning, the hippocampus is still necessary to learn contextual associations; in addition, the hippocampus is recruited to maintain an association between temporally dissociated events (Knight et al., 2004; McEchron et al., 1998; McEchron and Disterhoft, 1999). As suggested by Weitemier and Ryabinin (2003), the requirement of the hippocampus to keep the CS in short-term memory until the US presentation may alter hippocampal processing of contextual information. If this is the case, then there should be differences in hippocampal activity in trace versus standard fear conditioning. In support, studies have shown that trace cued fear conditioning develops after contextual fear conditioning (Barnet and Hunt, 2005; Moye and Rudy, 1987), and that there is differential activity of hippocampal cholinergic, glutamatergic, and GABAergic systems in trace versus standard fear conditioning (Hunt and Richardson, 2007; Wanisch et al., 2005; Wiltgen et al., 2005). Further research is needed to identify if these changes are responsible for the differential effects of ethanol on contextual learning in trace versus standard fear conditioning paradigms.

Passive Avoidance

Compared to fear conditioning, fewer studies have examined the acute effects of ethanol on passive avoidance. Nonetheless, those studies that have been conducted demonstrate that there are similar impairing effects of acute ethanol on fear conditioning and passive avoidance. These impairing effects may depend upon when ethanol is administered; administration before training may have different effects than administration after training or before recall. Studies have consistently shown that ethanol administered before training impairs passive avoidance learning in both rats and mice (Bammer and Chesher, 1982; Holloway, 1972; Rezayof et al., 2007). These results suggest that ethanol may impair acquisition, but do not rule out the possibility that ethanol may be altering consolidation of passive avoidance; furthermore, ethanol administration before training does not speak to the effects of ethanol on recall.

Unlike the results found with pre-training administration of ethanol, variable effects of ethanol on passive avoidance have been demonstrated following post-training administration. Alkana and Parker (1979) found that post-training administration of an acquisition-impairing dose of ethanol actually enhanced passive avoidance. However, Castellano and Pavone (1988) found that a dose range of ethanol similar to that used in the Alkana and Parker study impaired passive avoidance when administered post-training. Finally, Naylor and colleagues (2001) found no effect of post-training ethanol on passive avoidance conditioning, even though they also used a similar dose range. The differences between these studies may be due to differences in shock level and length of presentation of both the CS and the US. However, changes in shock intensity within the range used by these studies do not have a significant effect on conditioning (Bammer and Chesher, 1982), and all three studies saw similar levels of conditioning in controls, suggesting that there may be some as-yet unidentified explanation for the differential effects of post-training ethanol on passive avoidance in these studies. One possibility is that environmental factors may influence the effects of ethanol on passive avoidance. For example, Colbern and colleagues (1986) found that post-training administration of ethanol enhanced training if animals were returned to their home cage, but not if they were placed in a novel environment. Whereas it is clear that acute ethanol disrupts acquisition of passive avoidance, more research is needed on the effects of ethanol on consolidation of passive avoidance.

There is limited support for an effect of ethanol on recall of passive avoidance. Holloway (1972) found greater impairing effects of ethanol (2.0 g/kg) administered before training, but also found some impairment in avoidance behavior when ethanol was administered before testing. The testing day effect of ethanol may be explained in terms of anxiolysis, which could diminish fear and, thus, decrease the freezing response. This suggests that ethanol may disrupt performance in the passive avoidance test without directly altering recall. Interestingly, Holloway (1972) also found less impairment when ethanol was administered on both days compared to administration on only one day, suggesting a role of state dependence in the effects of ethanol on passive avoidance. In addition, state-dependent effects are seen in passive avoidance for lower doses of ethanol; a moderate dose of 0.5 g/kg ethanol impaired learning when administered before training but not when administered before training and testing (Rezayof et al., 2007). Nonetheless, Bammer and Chesher (1982) found no such state-dependency in passive avoidance conditioning using the same dose range as in the aforementioned studies. Thus, although state-dependency may play a role in the effects of ethanol on memory, ethanol also produces robust deficits in the acquisition of learning.

Factors Influencing the Effects of Ethanol on Aversive Learning

The effects of ethanol on aversive learning depend upon both the developmental state and the genetic background of the animal model being studied. There are differences in the sensitivity to ethanol between adolescent and adult animals, suggesting that there may be developmental changes in the way that the brain responds to ethanol treatment. Adolescent rats are less sensitive than adult rats to many of the impairing effects of ethanol (Silveri and Spear, 1998). For example, Land and Spear (2004) have demonstrated that adolescent rats were less impaired by ethanol in acquisition of both contextual and cued fear conditioning compared to adults, despite comparable blood alcohol levels across groups. Similarly, Hefner and Holmes (2006) found that adult mice were more susceptible to the effects of ethanol on contextual and cued fear conditioning compared to early adolescent and peri-adolescent mice. Thus, it appears that sensitivity to the effects of ethanol on learning increases with age. Reduced aversive effects of ethanol during adolescence may increase consumption during this period of high susceptibility to substance abuse (Carpenter-Hyland and Chandler, 2007; Crews et al., 2007).

Genetics are another factor that has been demonstrated to influence the effects of ethanol on multiple behaviors, including learning. There are many inter-strain differences in behavioral responses to ethanol treatment and understanding these genetic differences may aid in identifying risk factors in the development of alcoholism. For instance, different inbred strains of mice show different degrees of sensitivity, and even converse behavioral responses, to the effects of ethanol on locomotion (Crabbe et al., 1994; Crabbe et al., 2006; Crawley et al., 1997); some mouse strains, such as SWR/J, showed enhanced locomotor activity in response to 2 mg/kg ethanol, while other strains, such as AKR/J, showed depressed locomotion in response to the same dose (Crabbe et al., 1994). The influence of genetics is not limited to the effects of ethanol on locomotion but is also seen in the effects of ethanol on learning. Shapiro and Riley (1980) demonstrated that mice bred selectively for differential sensitivity to the motor-impairing effects of ethanol also showed differential levels of passive avoidance conditioning. Specifically, animals bred for high sensitivity to the effects of ethanol on locomotion showed better acquisition of avoidance learning (Shapiro and Riley, 1980). These results suggest a genetic correlation between sensitivity to the effects of ethanol on locomotion and aversive learning. Thus, studies examining the influences of genetics and development on ethanol-induced changes in learning will aid in identifying critical risk factors in the effects of alcohol on learning-related processes.

Summary

The combined research into the effects of acute ethanol on fear conditioning and passive avoidance has demonstrated robust impairments in acquisition of aversive learning, as well as some less-clear effects when ethanol is administered after training that may reflect changes in consolidation. The effects of acute ethanol on learning are complex, as task parameters, dose, developmental state, and genetics influence outcomes. In addition, it is still unclear if acute ethanol similarly affects fear conditioning and passive avoidance, as no studies have directly compared the effects of ethanol on these tasks. Further work is needed to identify whether there are differential effects of ethanol on these tasks and on the underlying neural substrates.

Effects of Chronic Ethanol and Withdrawal from Chronic Ethanol on Aversive Learning

As discussed, acute ethanol disrupts learning. One important issue is whether the same effects are seen across acute ethanol administration, chronic ethanol administration, and withdrawal from chronic ethanol. If acute, chronic, and withdrawal from chronic ethanol differentially affect learning, this could suggest that these states of ethanol administration are also differentially altering the underlying neural substrates of learning, and understanding these differences will facilitate development of treatments for alcohol abuse.

However, when comparing studies using chronic ethanol and withdrawal from chronic ethanol protocols, it is important to compare studies that examine similar time-points in ethanol treatment.

For example, some studies examining the effects of chronic ethanol on learning have tested animals during chronic administration, while others have withdrawn animals from ethanol and then tested them either immediately after withdrawal or after enough time has passed for withdrawal symptoms to have diminished.

These different time-points, during and after chronic administration, may differentially affect both the behavior and the neural substrates of the behavior. Indeed, it has been shown that chronic ethanol induces changes in NMDA and GABA receptor subunit expression, but most ethanol-induced changes in protein synthesis disappeared after withdrawal (Sheela Rani and Ticku, 2006), suggesting that the effects of ethanol after withdrawal may differ from those of chronic ethanol treatment.

If the molecular changes that underlie the effects of chronic ethanol return to baseline levels after withdrawal, then it is essential to study the effects of chronic ethanol treatment during the drug administration phase separately from the post-withdrawal effects of ethanol. Thus, in this review, studies examining the chronic effects of ethanol will be discussed in terms of when testing occurs – during chronic administration, during the initial withdrawal period, after multiple withdrawals, or after a long period of abstinence. Because of this, when talking about chronic treatment, we will describe both the dose and treatment duration.

EFFECTS OF CHRONIC ETHANOL ON AVERSIVE LEARNING

Passive Avoidance

Despite the different lengths of ethanol administration (1-6 months) used, there is a general impairment in passive avoidance induced by chronic ethanol when training and testing occur during ethanol administration. Markel and colleagues (1986) found that 28 days of chronic ethanol (~1.7 g/kg/day) consumption resulted in similar impairments of passive avoidance when conditioning occurred during day 10, 20, or 28 of chronic ethanol consumption.

Interestingly, they also found an increase in pain sensitivity with chronic ethanol treatment, as opposed to the decreased sensitivity found with acute treatment.

Logically, increased sensitivity to shock should increase conditioning in a passive avoidance task (Bammer and Chesher, 1982), so the persistent impairments in passive avoidance in the studies discussed herein suggest that the impairing effects of chronic ethanol are robust. Conversely, Sasaki and colleagues (1995) found no effect of ethanol (~8 mL/day of 15% EtOH) on passive avoidance after 20 weeks of chronic ethanol consumption. However, methodological differences such as shock level and dose of ethanol compared to the other studies discussed in this review could account for differences between studies. Farr and colleagues (2005) used comparable shock levels to those used in other studies and found that 8, but not 4, weeks of chronic ethanol (20% by volume) consumption impaired passive avoidance, suggesting that there is a minimum time for the development of chronic effects of ethanol on cognition. Thus, whereas both acute and chronic ethanol disrupt passive avoidance, the acute effects and the chronic effects of ethanol on passive avoidance may differ because deficits do not appear immediately in the chronic treatment paradigm.

Whereas the behavioral effects of chronic ethanol treatment on passive avoidance may not emerge immediately, some changes in underlying cellular and molecular processes may begin early in treatment and, over time, may become sufficient to disrupt behavior.

For example, Nixon and Crews (2002) have demonstrated that a 4-day binge ethanol (~9.3 g/kg/day) protocol is enough to decrease the survival of neural progenitor cells in the hippocampus, but changes in receptor density only occur after 6 weeks or longer of chronic ethanol administration (for review, Matthews and Morrow, 2000). Similar chronic ethanol-induced changes may occur in additional brain areas. Santucci and colleagues (2004) showed that 26 days of chronic ethanol (~16 g/kg/day) administration produced a decrease in cortical volume that persisted for 6 months after cessation of ethanol treatment; although this does not address the time-point at which the chronic effects of ethanol develop, it does suggest that chronic ethanol alters multiple neural substrates.

Fear Conditioning

Although many studies have examined the effects of chronic ethanol on passive avoidance, the effects of chronic ethanol on fear conditioning have yet to be examined; thus, task-dependent differences in the cognitive effects of chronic ethanol cannot be examined within the scope of the current review.

Effects of Withdrawal from Chronic Ethanol on Aversive Learning

Studies of the effects of withdrawal from chronic ethanol on aversive learning have used a range of paradigms that may model different aspects of ethanol withdrawal. For example, some studies have used a paradigm in which chronic ethanol is followed by a single withdrawal episode, while other studies have used repeated withdrawal episodes, and a third group of studies have examined the effects of withdrawal from ethanol after withdrawal symptoms have dissipated. A single withdrawal episode may model the effects of ethanol in abstaining alcoholics, while repeated withdrawal episodes during chronic administration may model the effects of binge drinking. Furthermore, changes in learning that remain after withdrawal symptoms dissipate may reflect long-lasting deficits and changes in underlying neural substrates. Thus, the remainder of this review will examine the short-term effects of withdrawal from chronic intermittent ethanol exposure and chronic continuous ethanol exposure, and the long-term effects of withdrawal from chronic ethanol.

Chronic Intermittent Ethanol and Withdrawal

Fear Conditioning

Fear conditioning has been shown to be particularly sensitive to repeated withdrawal episodes. Repeated withdrawal episodes (30 days of ~17.5 g/kg/day, with two 3-day withdrawals, starting at days 11 and 21) produced task-dependent impairments in fear conditioning (Ripley et al., 2003; Stephens et al., 2001) when conditioning occurred 12 days after the last withdrawal from ethanol, although a single withdrawal episode had no significant effect on fear conditioning at this time-point. Furthermore, fear conditioning was also impaired in human binge drinkers (Stephens et al., 2005). Interestingly, repeated withdrawal-induced deficits were specific for cued fear conditioning but not contextual fear conditioning (Borlikova et al., 2006; Ripley et al., 2003; Stephens et al., 2001; Stephens et al., 2005). In addition, associations that were formed before ethanol administration were not impaired by repeated withdrawal episodes, but new learning was impaired (Ripley et al.,

2003). Whereas it is unclear why repeated episodes, but not a single episode, disrupt cued fear conditioning, these repeated withdrawal episodes may differentially alter brain areas involved in cued versus contextual fear conditioning. Borlikova and colleagues (2006) suggest that the behavioral changes seen after withdrawal from chronic intermittent ethanol may be due to changes in cortical activity, but it remains to be determined if these changes are responsible for the differential effects of ethanol on cued versus contextual fear conditioning as well. Thus, whereas acute ethanol impairs both contextual and cued fear conditioning, multiple withdrawal experiences impair only cued fear conditioning, suggesting that these ethanol treatments may differentially impact the neural substrates involved in fear conditioning.

Passive Avoidance

In addition to having selective effects on cued fear conditioning, repeated withdrawal episodes did not alter passive avoidance (Celik et al., 2005; Popovic et al., 2004; Snell and Harris, 1979). This is in contrast to the acute effects of ethanol on passive avoidance, in which learning is disrupted. The results of these studies demonstrate that the effects of withdrawal from chronic intermittent ethanol on contextual fear conditioning, cued fear conditioning, and passive avoidance are distinct across tasks and differ from the acute effects of ethanol.

Effects of Single Withdrawal from Chronic Ethanol

Passive Avoidance

The effects of a single withdrawal episode on passive avoidance and fear conditioning are different than the effects of repeated withdrawal episodes. However, the time-course for the expression of deficits due to a single withdrawal from chronic ethanol varies across studies, making comparisons difficult. Celik and colleagues (2005) found impairments in passive avoidance 72 hours after withdrawal from 28 days of chronic ethanol (~13 g/kg/day). However, they found no impairment in passive avoidance at an early (24 hr) stage of withdrawal. The lack of withdrawal effects at 24 hours may be due to a delay in the molecular changes that occur in the transition from chronic effects to withdrawal effects. However, Snell and Harris (1979) found that withdrawal from 5 days of chronic oral ethanol administration (5% by volume) was enough to impair passive avoidance at 5 hours and at 5 days, but not at 2 weeks, post-withdrawal. In addition, animals on a chronic ethanol (~15g/kg/day) diet for 9 weeks, followed by 12 hours of withdrawal, were impaired in passive avoidance (Kuziemka-Leska et al., 1999). Thus, all three studies demonstrated that a single withdrawal episode disrupts passive avoidance; however, further work is needed to clarify why there are differences in the time-points at which the deficits occur. Finally, single withdrawal episodes differentially affect passive avoidance and fear conditioning, as a single episode was not enough to impair fear conditioning (Stephens et al., 2001); however, conditioning was only examined 12 days post-withdrawal, and thus more time-points for fear conditioning should be examined.

Post-Withdrawal Effects of Ethanol on Aversive Learning

As already discussed, many studies that use a chronic ethanol treatment paradigm actually test after the somatic signs of withdrawal have disappeared. The chronic effects of ethanol may persist after withdrawal, but molecular evidence argues against this possibility, as most chronic ethanol-induced changes in levels of glutamate and GABA receptor subunit mRNA returned to baseline levels after withdrawal (Sheela Rani and Ticku, 2006). Furthermore, it is unclear whether the cognitive and somatic symptoms of withdrawal persist for an equal length of time, and testing after the cessation of somatic withdrawal signs may not actually reflect the post-withdrawal effects of ethanol on cognition. Thus, the post-withdrawal effects of ethanol are discussed here as independent from the chronic effects of ethanol.

Fear Conditioning

Only one study has examined the post-withdrawal effects of ethanol on fear conditioning. Celerier and colleagues (2000) demonstrated that chronic oral ethanol consumption (12% by volume) severely impaired contextual fear conditioning without altering cued fear conditioning 4 weeks after cessation of an 11-month treatment. This is in sharp contrast to acute ethanol administration, in which contextual and cued conditioning were both sensitive to the impairing effects of ethanol (Gould, 2003; Gulick and Gould, 2007), and to the results from studies looking at multiple withdrawal episodes, which impaired only cued conditioning (Borlikova et al., 2006; Ripley et al., 2003; Stephens et al., 2001). One possible explanation for the dissociable effects of immediate withdrawal and long-term withdrawal from ethanol on fear conditioning may be the length of chronic treatment. The Celerier study (2000) administered ethanol for 11 months, while the Stephens study (2001), as aforementioned, administered ethanol for only one month. In addition, it is possible that long-term compensatory changes develop slowly in the neural substrates of learning that are altered by ethanol, producing behavioral effects long after the initial withdrawal episode. However, this possibility requires further research.

Passive Avoidance

The results from studies that examined the effects of chronic ethanol on passive avoidance after somatic withdrawal symptoms had disappeared (i.e. after at least one week of withdrawal from chronic ethanol) show variable results. Two studies have found deficits in passive avoidance after an extended period of withdrawal from chronic ethanol treatment. Animals that consumed ethanol (20% by volume) for 3 or 6 months, followed by 4 weeks without ethanol, were impaired in passive avoidance, although there was some recovery in the 3-month ethanol group (Casamenti et al., 1993). Similarly, Melis and colleagues (1996) withdrew rats after 36 weeks of chronic ethanol (~9 g/kg/day) consumption and then tested them in passive avoidance 15 days later and found a significant impairment of passive avoidance in ethanol-treated rats. However, Fadda and colleagues (1999) used a paradigm that achieved similar blood alcohol levels as Melis and colleagues (1996), and found no impairment in passive avoidance after ethanol treatment. Differences in methodology may explain the variability of these results. For example, a major difference between these two studies is that Melis and colleagues used a forced-consumption paradigm (ethanol was the only liquid available) while Fadda and colleagues used a drink-choice paradigm (ethanol and water were both available). Thus, it may be that consumption choice plays a role in the effects of chronic ethanol treatment on learning. In fact, studies have shown that passive administration and self-administration produced different drug effects (Donny et al., 2000; Greenwald and Roehrs, 2005). However, Santucci and colleagues (2004) used a forced-consumption paradigm to administer 26 days of ethanol (~16 g/kg/day) and found no effect of ethanol treatment on passive avoidance at 17 days after withdrawal. This suggests that drink choice may not be the factor contributing to the differences in the results between the Melis (1996) and Fadda (1999) studies. Finally, as discussed earlier, Snell and Harris (1979) administered 5 days of chronic ethanol (5% by volume) and found no effect at two weeks after withdrawal despite seeing an effect at earlier time-points. Both the duration of ethanol treatment and the time-point after treatment at which testing occurs may influence the effects of long-term withdrawal from ethanol on cognitive processes, but more research is needed to identify how these factors contribute to ethanol-induced deficits in learning. Furthermore, because deficits in passive avoidance are seen with immediate withdrawal and long-term withdrawal, it remains to be determined if the deficits found after an extended withdrawal are a continuation of the same deficits associated with immediate withdrawal, or if they represent unique deficits.

Cellular Processes Underlying Ethanol-induced Memory Deficits

As exemplified by results from behavioral studies, ethanol has complex effects on learning and memory. These complexities may reflect the many cellular and molecular substrates altered by ethanol. Thus, one way to further understand the behavioral effects of ethanol on learning and memory is to examine cellular processes of learning and memory that are altered by acute, chronic, and withdrawal from chronic ethanol treatment. Three neurotransmitters commonly implicated in learning and memory processes (for review, Myhrer, 2003) that may also be involved in the effects of ethanol on learning are glutamate, gamma-aminobutyric acid (GABA), and acetylcholine (Beracochea et al., 1986; Peris et al., 1997a; Schummers and Browning, 2001). Glutamate is the primary excitatory neurotransmitter in the brain and GABA is the primary inhibitory neurotransmitter in the brain. Unlike GABA and glutamate, acetylcholine is a modulatory neurotransmitter that can alter signaling in both GABAergic and glutamatergic systems (Ge and Dani, 2005; Yamazaki et al., 2005). As mentioned, these neurotransmitter systems underlie learning in the hippocampus and related structures (Myhrer, 2003). Furthermore, all three neurotransmitters are involved in the behavioral effects of acute and chronic ethanol (Nevo and Hamon, 1995).

As reviewed, acute ethanol disrupts both fear conditioning and passive avoidance (Bammer and Chesher, 1982; Gould, 2003; Melis et al., 1996). It is possible that changes in GABAergic, glutamatergic, and/or acetylcholinergic signaling modulate these effects. Administration of GABA receptor agonists disrupts both fear conditioning (Muller et al., 1997) and passive avoidance (Tohyama et al., 1991), and administration of glutamate antagonists also disrupts fear conditioning (Gould et al., 2002) and passive avoidance (Jafari-Sabet, 2006). Finally, cholinergic antagonists impair contextual fear conditioning (Gale et al., 2001; Vago and Kesner, 2007) and passive avoidance (Riekkinen et al., 1993). Acute ethanol potentiates GABA receptor-mediated activity (Suzdak and Paul, 1987) and inhibits glutamate receptor-mediated activity (Woodward et al., 2006; Yang et al., 1996); the summative effect of ethanol on these receptors is a depression of neural activity (Schummers et al., 1997; Schummers and Browning, 2001). Ethanol can also potentiate acetylcholinergic signaling (Forman and Zhou, 1999), which then alters the release of both GABA and glutamate (Wonnacott, 1997). Thus, ethanol may act on these neurotransmitter systems to alter learning. GABA antagonists block the impairing effects of ethanol on passive avoidance (Castellano and Pavone, 1988), while deletion of specific acetylcholine receptor subunits can decrease sensitivity to the memory-impairing effects of ethanol (Wehner et al., 2004). In addition, co-administration of ethanol and a sub-threshold dose of the glutamate antagonist MK-801 enhances the impairing effects of ethanol on passive avoidance (Aversano et al., 2002). This impairment is reversed by administration of GABA antagonists, suggesting that ethanol may facilitate interactions between GABAergic and glutamatergic processes. Further research is needed to compare the interactive effects of acute ethanol and all three neurotransmitter systems on fear conditioning and passive avoidance.

Like acute ethanol treatment, chronic ethanol alters learning. Whereas acute ethanol may modulate neurotransmitter receptor function, chronic ethanol produces changes in receptor density as well as receptor function. For example, chronic ethanol inhibits glutamate receptors, producing both a decrease in activity and a subsequent up-regulation of these receptors (Grant et al., 1990; Sanna et al., 1993; Snell et al., 1993). Because glutamate is critically involved in learning (Gould et al., 2002; Jafari-Sabet, 2006), perturbation of this system by chronic ethanol could contribute to learning deficits. Chronic ethanol also increases GABA receptor activity, leading to receptor desensitization and tolerance, and alters the subtypes of GABA receptors that are expressed (Grobin et al., 1998; Mhatre and Ticku, 1992; Sanna et al., 1993). Since these receptor subtypes have differential roles in cognition (Rudolph and Mohler, 2006), ethanol-induced changes in expression may also alter learning. In addition, chronic ethanol-associated neural changes, such as a decrease in GABA receptor sensitivity (Littleton and Little, 1994; Samson and Harris, 1992), that are associated with behavioral and cellular measures of tolerance (Allan and Harris, 1987; Grobin et al., 1998) may also contribute to ethanol-associated changes in learning-related processes. Finally, chronic ethanol decreases cholinergic function in many brain areas, and these changes have been linked to memory deficits associated with chronic ethanol consumption in multiple learning tasks (Arendt et al., 1989; Beracochea et al., 1986), although no studies have yet examined if the same changes alter fear conditioning or passive avoidance.

There are differences in the neural changes associated with chronic ethanol consumption and chronic intermittent ethanol consumption, just as there are differences in the behavioral effects of these treatments. Chronic and chronic intermittent ethanol produce differential up- and down-regulation of GABA and glutamate receptor subunits (Sheela Rani and Ticku, 2006), and these effects may be associated with different neural changes. For example, chronic intermittent ethanol compared to chronic ethanol has less of an effect on GABA receptor subunit expression but a greater effect on some glutamate receptor subunits (for review, Faingold et al., 1998; Sheela Rani and Ticku, 2006). Nonetheless, some GABA receptors are still up-regulated by chronic intermittent ethanol and GABA release is increased in the hippocampus (Peris et al., 1997b), which could produce changes in learning. It is unclear, however, if similar changes occur in other brain regions. Although chronic intermittent ethanol disrupts fear conditioning, it disrupts hippocampus-independent cued fear conditioning but not hippocampus-dependent passive avoidance and contextual fear conditioning (Popovic et al., 2004; Roberto et al., 2002; Stephens et al., 2001). Thus, the actions of chronic intermittent ethanol in other brain regions may underlie the behavioral changes associated with this treatment. Further examination of how chronic intermittent ethanol alters neurotransmitter release in brain regions other than the hippocampus will be essential in understanding the effects of ethanol on learning.

Withdrawal from chronic ethanol produces learning and memory deficits in fear conditioning and passive avoidance (Casamenti et al., 2003; Stephens et al., 2001). These deficits associated with withdrawal from chronic ethanol may be due to residual neural changes that are unmasked when chronic ethanol treatment ceases. For example, glutamate receptors increase in density but decrease in activation during chronic ethanol treatment; once ethanol is removed, the receptors rapidly return to a higher activation level, but more time is required for down-regulation of receptors (Hendricson et al., 2007). In addition, GABA receptors have reduced efficacy following chronic ethanol treatment (Kang et al., 1996); as a result, GABAergic inhibition is decreased when ethanol is removed, although these effects may vary between brain regions (for review, Weiner and Valenzuela, 2006). Both of these changes produce neuronal hyperexcitability, and this hyperexcitability is associated with the neural and behavioral symptoms of ethanol withdrawal (Little, 1999; Grant et al., 1990). Hyperexcitability produces indiscriminate neural activation; since learning requires the activation of specific neural networks, this over-activation may disrupt learning. Withdrawal from chronic ethanol also decreases acetylcholinergic function, and this decrease is associated with long-lasting deficits in passive avoidance (Casamenti et al., 1993), although it is unclear whether this deficit represents persistence of the chronic ethanol-induced deficit, or whether it represents a unique effect of ethanol withdrawal. Together, the studies reviewed suggest that the effects of acute ethanol, chronic ethanol, and withdrawal from chronic ethanol on glutamate-, GABA-, and acetylcholine-mediated processes may contribute to the ethanol-induced deficits in fear conditioning and passive avoidance, but more work is needed to link the variable effects of ethanol treatments on different types of learning to specific neural processes.

CONCLUSION

Alcohol abuse is a major societal issue, and one problem associated with acute and chronic alcohol consumption is disrupted cognition. Nonetheless, our understanding of the effects of ethanol on cognitive processes remains incomplete. Developments in the field of alcohol research have demonstrated that, contrary to initial theories, ethanol does not produce global deficits; rather, ethanol acts on specific substrates to alter neural function. The studies reviewed herein offer evidence for the specificity of the effects of ethanol on learning-related processes. These effects vary not only across learning tasks, but also vary within tasks with the phase of ethanol administration. Acute ethanol impairs both standard contextual and cued fear conditioning, as well as trace fear conditioning and passive avoidance; whereas the effects of chronic ethanol on fear conditioning are unknown, chronic ethanol also impairs passive avoidance. The effects of acute ethanol on passive avoidance may differ from the effects of chronic ethanol because shorter chronic treatments do not disrupt passive avoidance but extended treatments do. Withdrawal from chronic ethanol also disrupts learning. A single immediate withdrawal episode impairs passive avoidance without altering fear conditioning, although only one time-point was examined, while multiple withdrawal episodes impair cued fear conditioning without altering contextual fear conditioning or passive avoidance. Finally, long-term withdrawal from ethanol impairs contextual fear conditioning but leaves cued fear conditioning intact; the effects of long-term withdrawal on passive avoidance are variable and depend on the protocol being employed (Table 1). Thus, the duration of ethanol treatment and the duration of withdrawal from chronic treatment may differentially influence learning and underlying neural substrates. In addition, as reviewed, other factors such as genetics and age can influence the effects of ethanol on learning-related processes even within a single task. Understanding how genetics, age, and duration of ethanol treatment interact to alter both learning and its underlying neural substrates will be important for understanding and treating alcohol abuse and alcoholism.

REFERENCES

Abe, K. and Misawa, M. GABAA receptor-mediated inhibition by ethanol of long-term potentiation in the basolateral amygdala-dentate gyrus pathway in vivo. *Neuroscience*, 125(1): 113-117, 2004.

Alkana, R.L. and Parker, E.S. Memory facilitation by post-training injection of ethanol. *Psychopharm*, 66(2): 117-119, 1979.

Allan, A.M. and Harris, R.A. Involvement of neuronal chloride channels in ethanol intoxication, tolerance, and dependence. *Rec. Dev. Alcohol* 5: 313-325. 1987.

Ambrogi Lorenzini, C., Bucherelli, C., Giachetti, A., Mugnai, L. and Tassoni, G. Effects of nucleus basolateralis amygdalae neurotoxic lesions on aversive conditioning in the rat. *Physiol. Behav.*, 49(4): 765-770, 1991.

Arendt, T., Allen, Y., Marchbanks, R.M., Schugens M.M., Sinden, J., Lantos, P.L. and Gray, J.A. Cholinergic system and memory in the rat: effects of chronic ethanol, embryonic basal forebrain brain transplants and excitotoxic lesions of cholinergic basal forebrain projection system. *Neuroscience* 33(3): 435-462, 1989.

Aversano, M., Ciamei, A., Cestari, V., Passino, E., Middei, S. and Castellano, C. Effects of MK-801 and ethanol combinations on memory consolidation in CD1 mice: involvement of GABAergic mechanisms. *Neurobio Learn Mem* 77(3): 327-337, 2002.

Bammer, G. and Chesher, G.B. An analysis of some effects of ethanol on performance in a passive avoidance task. *Psychopharm*, 77: 66-73, 1982.

Barnet, R.C. and Hunt, P.S. Trace and long-delay fear conditioning in the developing rat. *Learn Behav.*, 33(4): 437-443, 2005.

Beracochea, D., Durkin, T.P. and Jaffard, R. On the involvement of the central cholinergic system in memory deficits induced by long term ethanol consumption in mice. *Pharmacol. Biochem. Behav.* 24(3): 519-524, 1986.

Birnbaum, I.M., Parker, E.S., Hartley, J.T. and Noble, E.P. Alcohol and memory: retrieval processes. *J. Verb. Learn Verb. Behav.*, 17: 325-335, 1978.

Blanchard, R.J. and Blanchard, D.C. Aggressive behavior in the rat. *Behav. Biol.*, 21(2): 197-224, 1977.

Borlikova, G.G., Elbers, N.A. and Stephens, D.N. Repeated withdrawal from ethanol spares contextual fear conditioning and spatial learning but impairs negative patterning and induces over-responding: evidence for effect on frontal cortical but not hippocampal function? *Euro. J. Neurosci.*, 24(1): 205-216, 2006.

Buchel, C., Dolan, R.J., Armony, J.L. and Friston, K.J. Amygdala-hippocampal involvement in human aversive trace conditioning revealed through event-related functional magnetic resonance imaging. *J. Neurosci.* 19(24): 10869-76, 1999.

Carpenter-Hyland, E.P. and Chandler, L.J. Adaptive plasticity of NMDA receptors and dendritic spines: Implications for enhanced vulnerability of the adolescent brain to alcohol addiction. *Pharmacol. Biochem. Behav.* 86(2): 200-208, 2007.

Carter, R.M., Hofstotter, C., Tsuchiya, N. and Koch, C. Working memory and fear conditioning. *PNAS,* 100(3): 1399-1404, 2003.

Casamenti, F., Scali, C., Vannucchi, M.G., Bartolini, L. and Pepeu, G. Long-term ethanol consumption by rats: effect on acetylcholine release in vivo, choline acetyltransferase activity, and behavior. *Neurosci.,* 56(2): 465-471, 1993.

Castellano, C. and Pavone, F. Effects of ethanol on passive avoidance behavior in the mouse: involvement of GABAergic mechanisms. *Pharmacol. Biochem. Behav.*, 29(2): 321-324, 1988.

Celerier, A., Ognard, R., Decorte, L. and Beracochea, D. Deficits of spatial and non-spatial memory and of auditory fear conditioning following anterior thalamic lesions in mice: comparison with chronic alcohol consumption. *Euro. J. Neuro.*, 12: 2575-2584, 2000.

Celik, T., Cakir, E., Kayir, H., Bilgi, C. and Uzbay, I.T. The effects of chronic ethanol consumption and withdrawal on passive avoidance task and serum cholinesterase level in rats. *Prog Neuropsychopharmacol Biol. Psychiatry*, 29(4): 505-509, 2005.

Colbern, D.L., Sharek, P. and Zimmermann, E.G. The effect of home or novel environment on the facilitation of passive avoidance by post-training ethanol. *Behav. Neural Biol.*, 46(1): 1-12, 1986.

Crabbe, J.C. Where does alcohol act in the brain? *Mol. Psych.*, 2: 17-20, 1997.

Crabbe, J.C., Gallaher, E.S., Phillips, T.J., Belknap, J.K. Genetic determinants of sensitivity to ethanol in inbred mice. *Behav. Neurosci.*, 108(1): 186-95, 1994.

Crabbe, J.C., Phillips, T.J., Harris, R.S., Arends, M.A. and Koob, G.F. Alcohol-related genes: contributions from studies with genetically engineered mice. *Add Bio.*, 11(3-4): 195-269, 2006.

Crawley, J.N., Belknap, J.K., Collins, A., Crabbe, J.C., Frankel, W., Henderson, N., Hitzemann, R.J., Maxson, S.C., Miner, L.L., Silva, A.J., Wehner, J.M., Wynshaw-Boris, A. and Paylor, R. Behavioral phenotypes of inbred mouse strains: implications and recommendations for molecular studies. *Psychopharm.*, 132 (2): 107-124, 1997.

Crews, F., He, J. and Hodge, C. Adolescent cortical development: A critical period of vulnerability for addiction. *Pharmacol. Biochem. Behav.*, 86(2): 189-199, 2007.

Cunningham, C.L. and Noble, D. Conditioned activation induced by ethanol: role in sensitization and conditioned place preference. *Pharmacol Biochem. Behav.* 43(1): 307-13, 1992.

Cutler, R.B. and Fishbain, D.A. Are alcoholism treatments effective? The project MATCH data. *BMC Pub. Health*, 5: 75-86, 2005.

Dell'Omo, G. and Alleva, E. Snake odor alters behavior, but not pain sensitivity in mice. *Physiol. Behav.*, 55(1): 125-128, 1994.

Donny, E.C., Caggiula, A.R., Rose, C., Jacobs, K.S., Mielke, M.M. and Sved, A.F. Differential effects of response-contingent and response-independent nicotine in rats. *Eur. J. Pharmacol.*, 402(3): 231-240, 2000.

Fadda, F. and Rossetti, Z.L. Chronic ethanol consumption: from neuroadaptation to neurodegeneration. *Prog. in Neurobio*, 56: 385-431, 1998.

Fadda, F., Cocco, S., Stancampiano, R. and Rossetti, Z.L. Long-term voluntary ethanol consumption affects neither spatial nor passive avoidance learning, nor hippocampal acetylcholine release in alcohol-preferring rats. *Behav. Brain Res.*, 103(1): 71-76, 1999.

Faingold, C.L., N'Gouemo, P. and Riaz, A. Ethanol and n eurotransmitter interactions--from molecular to integrative effects. *Prog. Neurobiol.* 55(5): 509-535, 1998.

Fanselow, M.S. Contextual fear, gestalt memories, and the hippocampus. *Behav. Brain Res.*, 110(1-2): 73-81, 2000.

Farr, S.A., Scherrer, J.F., Banks, W.A., Flood, J.F. and Morley, J.E. Chronic ethanol consumption impairs learning and memory after cessation of ethanol. *Alc. Clin. Exp. Res.*, 29 (6): 971-982, 2005.

Faw B. Pre-frontal executive committee for perception, working memory, attention, long-term memory, motor control, and thinking: a tutorial review. *Conscious Cogn*, 12(1): 83-139, 2003.

Fendt, M. and Fanselow, M.S. The neuroanatomical and neurochemical basis of conditioned fear. *Neurosci. Biobehav. Rev.*, 23: 743-760, 1999.

Forman, S.A. and Zhou, Q. Novel modulation of a nicotinic receptor channel mutant reveals that the open state is stabilized by ethanol. *Mol. Pharm.*, 55: 102-108, 1999.

Gale, G.D., Anagnostaras, S.G. and Fanselow, M.S. Cholinergic modulation of pavlovian fear conditioning: effects of intrahippocampal scopolamine infusion. *Hippocampus* 11(4): 371-376, 2001.

Ge, S. and Dani, J.A. Nicotinic acetylcholine receptors at glutamate synapses facilitate long-tem depression or potentiation. *J. Neuro*, 25(26): 6084-6091, 2005.

Gessa, G.L., Muntoni, F., Collu, M., Vargiu, L. and Mereu, G. Low doses of ethanol activate dopaminergic neurons in the ventral tegmental area. *Brain Res.*, 348: 201-203, 1985.

Givens, B. Low doses of ethanol impair spatial working memory and reduce hippocampal theta activity. *Alc. Clin. Exp. Res.*, 19 (3): 763-767, 1995.

Givens, B. and McMahon, K. Effects of ethanol on nonspatial working memory and attention in rats. *Behav. Neuro.*, 111(2): 275-282, 1997.

Gould, T.J. Ethanol disrupts fear conditioning in C57BL/6J mice. *J. Psychopharm.*, 17(1): 77-81, 2003.

Gould, T.J., McCarthy, M.M. and Keith, R.A. MK-801 disrupts acquisition of contextual fear conditioning but enhances memory consolidation of cued fear conditioning. *Behav. Pharmacol.* 13(4): 287-294, 2002.

Gould, T.J. and Lommock, J.A. Nicotine enhances contextual fear conditioning and ameliorates ethanol-induced deficits in contextual fear conditioning. *Behav. Neuro.*, 117 (6): 1276-1282, 2003.

Grant, K.A., Valverius, P., Hudspith, M. and Tabakoff, B. Ethanol withdrawal seizures and the NMDA receptor complex. *Eur. J. Pharm.*, 176: 289-296, 1990.

Greenwald, M.K. and Roehrs, T.A. Mu-opioid self-administration vs. passive administration in heroin abusers produces differential EEG activation. *Neuropsychpharm*, 30(1): 212-221, 2005.

Grobin, A.C., Matthews, D.B., Devaud, L.L. and Morrow, A.L. The role of GABA(A) receptors in the acute and chronic effects of ethanol. *Psychopharm* 139(1-2): 2-19, 1998.

Gulick, D. and Gould, T.J. Acute ethanol has biphasic effects on short- and long-term memory in both foreground and background contextual fear conditioning in C57BL/6 mice. *Alc. Clin. Exp. Res.*, 31(9): 1528-1537, 2007.

Gulick D. and Gould T.J. Interactive effects of ethanol and nicotine on learning in C57BL/6J mice depend on both dose and duration of treatment. *Psychopharm,* 196(3): 483-495, 2008a.

Gulick, D. and Gould, T.J. Varenicline Ameliorates Ethanol-Induced Deficits in Learning in C57BL/6 Mice. *Neurobio Learn Mem.,* in press.

Hefner K. and Holmes A. An investigation of the behavioral actions of ethanol across adolescence in mice. *Psychopharm*, 191(2): 311-322, 2006.

Heinz, A., Siessmeier, T., Wrase, J., Hermann, D., Klein, S., Grusser, S.M., Flor, H., Braus, D.F., Buchholz, H.G., Grunder, G., Schreckenberger, M., Smolka, M.N., Rosch, F., Mann, K. and Bartenstein, P. Correlation between dopamine D(2) receptors in the ventral striatum and central processing of alcohol cues and craving. *Am. J. Psychiatry*, 161(10): 1783-1789, 2004.

Hendricson, A.W., Maldve, R.E., Salinas, A.G., Theile, J.W., Zhang, T.A., Diaz, L.M. and Morrisett, R.A. Aberrant synaptic activation of N-methyl-D-aspartate receptors underlies ethanol withdrawal hyperexcitability. *J. Pharmacol. Exp. Ther.*, 321(1): 60-72, 2007.

Hernandez, L.L. and Powell, D.A. Ethanol enhancement of Pavlovian conditioning: comparison with instrumental conditioning. *Psychopharm*, 88: 75-81, 1986.

Hernandez, L.L., Valentine, J.D. and Powell, D.A. Ethanol enhancement of Pavlovian conditioning. *Behav. Neurosci.,* 100(4): 494-503, 1986.

Hernandez, L.L. and Valentine, J.D. Mild ethanol intoxication may enhance pavlovian conditioning. *Drug Dev. Res.*, 20: 155-167, 1990.

Higgins, S.T., Rush, C.R., Hughes, J.R., Bickel, W.K., Lynn, M. and Capeless, M.A. Effects of cocaine and alcohol, alone and in combination, on human learning and performance. *J. Exp. An. Behav.*, 58 (1): 87-105, 1992.

Himwich, H.E. The physiological action of alcohol. Emerson H. (Ed.), Alcohol and Man, 1-23, The MacMillan Company, New York, 1932.

Holloway, F.A. State-dependent effects of ethanol on active and passive avoidance learning. *Psychopharm*, 25(3): 238-261, 1972.

Hunt, P.S. and Richardson, R. Pharmacological dissociation of trace and long-delay fear conditioning in young rats. *Neurobiol. Learn Mem.*, 87(1): 86-92, 2007.

Izquierdo, I., Bevilaqua, L.R., Rossato, J.I., Bonini, J.S., Da Silva, W.C., Medina, J.H. and Cammarota, M. The connection between the hippocampal and the striatal memory systems of the brain: a review of recent findings. *Neurotox Res.*, 10(2): 113-121, 2006.

Jafari-Sabet, M. NMDA receptor blockers prevents the facilitatory effects of post-training intra-dorsal hippocampal NMDA and physostigmine on memory retention of passive avoidance learning in rats. *Behav. Brain Res.* 169(1): 120-127, 2006.

Kang, M., Spigelman, I., Sapp, D.W. and Olsen, R.W. Persistent reduction of GABA(A) receptor-mediated inhibition in rat hippocampus after chronic intermittent ethanol treatment. *Brain Res.* 709(2): 221-228, 1996.

Knight, D.C., Cheng, D.T., Smith, C.N. and Stein, E.A., Helmstetter, F.J. Neural substrates mediating human delay and trace fear conditioning. *J. Neurosci.*, 24(1): 218-28, 2004.

Knight, D.C., Nguyen, H.T. and Bandettini, P.A. The role of awareness in delay and trace fear conditioning in humans. *Cogn Affect Behav. Neurosci.* 6(2): 157-62, 2006.

Knudsen, E.I. Fundamental components of attention. *Annu. Rev. Neurosci.* 30: 57-78, 2007.

Koob, G.F. A role for GABA mechanisms in the motivational effects of alcohol. *Biochem. Pharmacol.* 68(8):1515-25, 2004.

Korte, S.M. Corticosteroids in relation to fear, anxiety and psychopathology. *Neurosci. Biobehav. Rev.*, 25(2): 117-142, 2001.

Kuziemka-Leska, M., Car, H. and Wisniewski, K. Baclofen and AII 3-7 on learning and memory processes in rats chronically treated with ethanol. *Pharmacol. Biochem. Behav.*, 62(1): 39-43, 1999.

Land, C. and Spear, N.E. Fear conditioning is impaired in adult rats by ethanol doses that do not affect periadolescents. *Int. J. Dev. Neurosci.*, 22(5-6): 355-362, 2004.

LeDoux, J. The emotional brain, fear, and the amygdala. *Cell Mol. Neurobiol.*, 23(4-5): 727-38, 2003.

Lima-Landman, M.T. and Albuquerque, E.X. Ethanol potentiates and blocks NMDA-activated single-channel currents in rat hippocampal pyramidal cells. *FEBS Lett* 247(1): 61-67, 1989.

Little, H.J. The contribution of electrophysiology to knowledge of the acute and chronic effects of ethanol. *Pharmacol Ther,* 84(3): 333-353, 1999.

Littleton, J. and Little, H. Current concepts of ethanol dependence. *Addiction* 89(11): 1397-1412, 1994.

Logue, S.F., Paylor, R. and Wehner, J.M. Hippocampal lesions cause learning deficits in inbred mice in the Morris water maze and conditioned-fear task. *Behav. Neurosci.,* 111(1): 104-133, 1997.

Lyon, R.J., Tong, J.E., Leigh, G. and Clare, G. The influence of alcohol and tobacco on the components of choice reaction time. *J. Stud. Alc.*, 36(5): 587-596, 1975.

Markel, E., Nyakas, C. and Tal, E., Endroczi, E. Changes in avoidance behaviour following ethanol treatment in rats of different ages. *Acta Physiol. Hung*, 68(2): 175-181, 1986.

Martinez, I., Quirarte, G.L., Diaz-Cintra, S., Quiroz, C. and Prado-Alcala, R.A. Effects of lesions of hippocampal fields CA1 and CA3 on acquisition of inhibitory avoidance. *Neuropsychobiol.*, 46(2): 97-103, 2002.

Matsushima, T., Izawa, E., Aoki, N. and Yanagihara, S. The mind through chick eyes: memory, cognition and anticipation. *Zoolog. Sci.*, 20(4): 395-408, 2003.

Matthews, D.B. and Morrow, A.L. Effects of acute and chronic ethanol exposure on spatial cognitive processing and hippocampal function in the rat. *Hippocampus* 10(1): 122-30, 2000.

McBride, W.J. Central nucleus of the amygdala and the effects of alcohol and alcohol-drinking behavior in rodents. *Pharmacol. Biochem. Behav.* 71(3): 509-15, 2002.

McEchron, M.D., Bouwmeester, H., Tseng, W., Weiss, C. and Disterhoft, J.F. Hippocampectomy disrupts auditory trace fear conditioning and contextual fear conditioning in the rat. *Hippocampus* 8(6): 638-46, 1998.

McEchron, M.D. and Disterhoft, J.F. Hippocampal encoding of non-spatial trace conditioning. *Hippocampus* 9(4): 385-96, 1999.

McNally, G.P. and Westbrook, R.F. Predicting danger: the nature, consequences, and neural mechanisms of predictive fear learning. *Learn Mem.*, 13(3): 245-253, 2006.

Melia, K.R., Ryabinin, A.E., Corodimas, K.P., Wilson, M.C. and LeDoux, J.E. Hippocampal-dependent learning and experience-dependent activation of the hippocampus are preferentially disrupted by ethanol. *Neurosci.*, 74 (2): 313-322, 1996.

Melis, F., Stancampiano, R., Imperato, A., Carta, G. and Fadda, F. Chronic ethanol consumption in rats: correlation between memory performance and hippocampal acetylcholine release in vivo. *Neuroscience*, 74(1): 155-159, 1996.

Mhatre, M.C. and Ticku, M.K. Chronic ethanol administration alters gamma-aminobutyric acidA receptor gene expression. *Mol. Pharmacol.,* 42(3): 415-422, 1992.

Moye, T.B. and Rudy, J.W. Ontogenesis of trace conditioning in young rats: dissociation of associative and memory processes. *Dev. Psychobiol.*, 20(4): 405-414, 1987.

Muller, J., Corodimas, K.P., Fridel, Z. and LeDoux, J.E. Functional inactivation of the lateral and basal nuclei of the amygdala by muscimol infusion prevents fear conditioning to an explicit conditioned stimulus and to contextual stimuli. *Behav. Neurosci.* 111(4): 683-691, 1997.

Myhrer, T. Neurotransmitter systems involved in learning and memory in the rat: a meta-analysis based on studies of four behavioral tasks. *Brain Res. Brain Res. Rev.* 41(2-3): 268-287, 2003.

Nader, K., Majidishad, P., Amorapanth, P. and LeDoux, J.E. Damage to the lateral and central, but not other, amygdaloid nuclei prevents the acquisition of auditory fear conditioning. *Learn Mem.* 8(3): 156-163, 2001.

National Institute on Alcohol Abuse and Alcoholism (2004): http://pubs.niaaa.nih.gov/publications/GettheFacts_HTML/Facts.pdf.

Naylor, J.C., Simson, P.E., Gibson, B., Schneider, A.M., Wilkins, E. and Firestone, A., Choy, M. Ethanol inhibits spontaneous activity of central nucleus of the amygdala neurons but does not impair retention in the passive-avoidance task. *Alc Clin. Exp. Res.*, 25(11): 1683-1688, 2001.

Nevo, I. and Hamon, M. Neurotransmitter and neuromodulatory mechanisms involved in alcohol abuse and alcoholism. *Neurochem. Int.* 26(4) 305-336, 1995.

Nixon, K. and Crews, F.T. Binge ethanol exposure decreases neurogenesis in adult rat hippocampus. *J. Neurochem.*, 83(5): 1087-93, 2002.

Peris, J., Anderson, K.J., Vickroy, T.W., King, M.A., Hunter, B.E. and Walker, D.W. Neurochemical basis of disruption of hippocampal long term potentiation by chronic alcohol exposure. *Front Biosci.*, 2: d309-16, 1997a.

Peris, J., Eppler, B., Hu, M., Walker, D.W., Hunter, B.E., Mason, K. and Anderson, K.J. Effects of chronic ethanol exposure on GABA receptors and GABAB receptor modulation of 3H-GABA release in the hippocampus. *Alc. Clin. Exp. Res.,* 21(6): 1047-1052, 1997b.

Phillips, R.G. and LeDoux, J.E. Differential contribution of amygdala and hippocampus to cued and contextual fear conditioning. *Behav. Neurosci.,* 106 (2): 274-285, 1992.

Pihl, R.O. and Peterson, J.B. Alcoholism: The role of different motivational systems. *J. Psych. Neurosci.*, 20(5): 372-397, 1995.

Popke, E.J., Allen, S.R. and Paule, M.G. Effects of acute ethanol on indices of cognitive-behavioral performance in rats. *Alcohol*, 20: 187-192, 2000.

Popovic, M., Caballero-Bleda, M., Puelles, L. and Guerri, C. Multiple binge alcohol consumption during rat adolescence increases anxiety but does not impair retention in the passive avoidance task. *Neurosci. Lett.*, 357(2): 79-82, 2004.

Reid, L.D., Hunter, G.A., Beaman, C.M. and Hubbell, C.L. Toward understanding ethanol's capacity to be reinforcing: a conditioned place preference following injections of ethanol. *Pharmacol. Biochem. Behav.*, 22(3): 483-7, 1985.

Rezayof, A., Motevasseli, T., Rassouli, Y. and Zarrindast, M.R. Dorsal hippocampal dopamine receptors are involved in mediating ethanol state-dependent memory. *Life Sci.,* 80(4): 285-292, 2007.

Rezvani, A.H. and Levin, E.D. Nicotine-alcohol interactions and attentional performance on an operant visual signal detection task in female rats. *Pharm. Biochem. Behav.*, 76: 75-83, 2003.

Riekkinen, P. Jr, Riekkinen, M. and Sirviö, J. Cholinergic drugs regulate passive avoidance performance via the amygdala. *J. Pharmacol. Exp. Ther.* 267(3): 1484-1492, 1993.

Ripley, T.L., O'Shea, M. and Stephens, D.N. Repeated withdrawal from ethanol impairs acquisition but not expression of conditioned fear. *Euro. J. Neuro.,* 18: 441-448, 2003.

Roberto, M., Bajo, M., Crawford, E., Madamba, S.G. and Siggins, G.R. Chronic ethanol exposure and protracted abstinence alter NMDA receptors in central amygdala. *Neuropsychopharm,* 31(5): 988-996, 2006.

Roberto, M., Nelson, T.E., Ur, C.L. and Gruol, D.L. Long-term potentiation in the rat hippocampus is reversibly depressed by chronic intermittent ethanol exposure. *J. Neurophysiol.,* 87(5): 2385-2397, 2002.

Rodrigues, S.M., Schafe, G.E. and LeDoux, J.E. Molecular mechanisms underlying emotional learning and memory in the lateral amygdala. *Neuron,* 44(1): 75-91, 2004.

Rudolph U. and Möhler H. GABA-based therapeutic approaches: GABAA receptor subtype functions. *Curr. Opin. Pharmacol.,* 6(1): 18-23, 2006.

Rudy, J.W., Barrientos, R.M., O'Reilly, R.C. Hippocampal formation supports conditioning to memory of a context. *Behav. Neurosci.*, 116(4): 530-538, 2002

Ryback, R.S. The continuum and specificity of the effects of alcohol on memory. *Q. J. Stud. Alc.,* 32: 996-1016, 1971.

Samson, H.H. and Harris, R.A. Neurobiology of alcohol abuse. *Trends Pharmacol. Sci.* 13(5): 206-211, 1992.

Sanna, E., Serra, M., Cossu, A., Colombo, G., Follesa, P., Cucceddu, T., Concas, A. and Biggio, G. Chronic ethanol intoxication induces differential effects on GABAA and NMDA receptor function in the rat brain. *Alc. Clin. Exp. Res.*, 17(1): 115-123, 1993.

Santucci, A.C., Mercado, M., Bettica, A., Cortes, C., York, D. and Moody, E. Residual behavioral and neuroanatomical effects of short-term chronic ethanol consumption in rats. *Brain Res. Cogn. Brain Res.*, 20(3): 449-461, 2004.

Sasaki, H., Matsuzaki, Y., Nakagawa, T., Arai, H., Yamama, M., Sekizawa, K., Ikarashi, Y. and Maruyama, Y. Cognitive function in rats with alcohol ingestion. *Pharmacol. Biochem. Behav.*, 52(4): 845-848, 1995.

Schummers, J., Bentz, S. and Browning, M.D. Ethanol's inhibition of LTP may not be mediated solely via direct effects on the NMDA receptor. *Alc. Clin. Exp. Res.*, 21(3): 404-408, 1997.

Schummers J. and Browning M.D. Evidence for a role for GABA(A) and NMDA receptors in ethanol inhibition of long-term potentiation. *Brain Res. Mol. Brain Res.* 94(1-2): 9-14, 2001.

Shapiro N.R. and Riley E.P. Avoidance behavior in rats selectively bred for differential alcohol sensitivity. *Psychopharm.*, 72(1): 79-83, 1980.

Sheela Rani, C.S. and Ticku, M.K. Comparison of chronic ethanol and chronic intermittent ethanol treatments on the expression of GABA(A) and NMDA receptor subunits. *Alcohol.* 38(2): 89-97, 2006.

Silveri, M.M. and Spear, L.P. Decreased sensitivity to the hypnotic effects of ethanol early in ontogeny. *Alc. Clin. Exp. Res.*, 22(3): 670-676, 1998.

Snell, L.D., Tabakoff, B. and Hoffman, P.L. Radioligand binding to the NMDA receptor/ionophore complex: alterations by ethanol in vitro and by chronic in vivo ethanol ingestion. *Brain Res.*, 602: 91-98, 1993.

Snell, D. and Harris, R.A. Impairment of avoidance behavior following short-term ingestion of alcohol. *Psychopharm*, 63(3): 251-257, 1979.

Stephens, D.N., Brown, G., Duka, T. and Ripley, T.L. Impaired fear conditioning but enhanced seizure sensitivity in rats given repeated experience of withdrawal from alcohol. *Euro. J. Neuro.*, 14: 2032-2031, 2001.

Stephens, D.N., Ripley, T.L., Borlikova, G., Schubert, M., Albrecht, D., Hogarth, L. and Duka, T. Repeated ethanol exposure and withdrawal impairs human fear conditioning and depresses long-term potentiation in rat amygdala and hippocampus. *Biol. Psych.*, 58: 392-400, 2005.

Suzdak, P.D. and Paul, S.M. Ethanol stimulates GABA receptor-mediated Cl- ion flux in vitro: possible relationship to the anxiolytic and intoxicating actions of alcohol. *Psychopharmacol. Bull*, 23(3): 445-451, 1987.

Tinsley, M.R., Quinn, J.J. and Fanselow M.S. The role of muscarinic and nicotinic cholinergic neurotransmission in aversive conditioning: comparing pavlovian fear conditioning and inhibitory avoidance. *Learn Mem.,* 11(1): 35-42, 2004.

Tohyama, K., Nabeshima, T., Ichihara, K. and Kameyama, T. Involvement of GABAergic systems in benzodiazepine-induced impairment of passive avoidance learning in mice. *Psychopharm* 105(1): 22-26, 1991.

Tokuda, K., Zorumski, C.F. and Izumi, Y. Modulation of hippocampal long-term potentiation by slow increases in ethanol concentration. *Neurosci.,* 146(1): 340-349, 2007.

Vago, D.R. and Kesner, R.P. Cholinergic modulation of Pavlovian fear conditioning in rats: differential effects of intrahippocampal infusion of mecamylamine and methyllycaconitine. *Neurobiol. Learn Mem.* 87(3): 441-449, 2007.

Vertes, R.P. Brain stem generation of the hippocampal EEG. *Prog. Neurobiol.* 19: 159-186, 1982.

Vogel, R.A., Frye, G.D., Wilson J.H., Kuhn C.M., Koepke K.M., Mailman R.B., Mueller R.A. and Breese G.R. Attenuation of the effects of punishment by ethanol: comparisons with chlordiazepoxide. *Psychopharm.*, 71(2): 123-129, 1980.

Wagner, A.F. and Hunt, P.S. Impaired trace fear conditioning following neonatal ethanol: reversal by choline. *Behav. Neurosci.,* 120(2): 482-487, 2006.

Wallace, G.B. The physiological action of alcohol. Emerson H. (Ed.), Alcohol and Man, 24-65, The MacMillan Company, New York, 1932.

Wanisch, K., Tang, J., Mederer A. and Wotjak C.T. Trace fear conditioning depends on NMDA receptor activation and protein synthesis within the dorsal hippocampus of mice. *Behav. Brain Res.*, 157(1): 63-69, 2005.

Wehner, J.M., Keller, J.J., Keller, A.B., Picciotto, M.R., Paylor, R., Booker, T.K., Beaudet, A., Heinemann, S.F. and Balogh, S.A. Role of neuronal nicotinic receptors in the effects of nicotine and ethanol on contextual fear conditioning. *Neuroscience* 129(1): 11-24, 2004.

Weiner, J.L. and Valenzuela, C.F. Ethanol modulation of GABAergic transmission: the view from the slice. *Pharmacol. Ther.* 111(3): 533-554, 2006.

Weitemier, A.E. and Ryabinin, A.E. Alcohol-induced memory impairment in trace fear conditioning: a hippocampus-specific effect. *Hippocampus*, 13: 305-315, 2003.

Wilensky, A.E., Schafe, G.E. and LeDoux J.E. The amygdala modulates memory consolidation of fear-motivated inhibitory avoidance learning but not classical fear conditioning. *J. Neurosci.*, 20(18): 7059-7066, 2000.

Wilkinson, D.A. and Poulos, C.X. The chronic effects of alcohol on memory: a contrast between a unitary and dual system approach. Galanter (Ed.), Rec Dev Alc. 5-26, Plenum Press, New York, 1987.

Wiltgen, B.J., Sanders, M.J., Ferguson C., Homanics G.E. and Fanselow M.S. Trace fear conditioning is enhanced in mice lacking the delta subunit of the GABAA receptor. *Learn Mem.*, 12(3): 327-333, 2005.

Wonnacott, S. Presynaptic nicotinic Ach receptors. *Trends Neurosci.*, 20(2): 92-98, 1997.

Woodward, J.J., Ron, D., Winder, D. and Roberto, M. From blue states to up states: a regional view of NMDA-ethanol interactions. *Alc. Clin. Exp. Res.*, 30(2): 359-367, 2006.

Yamazaki, Y., Jia, Y., Hamaue, N. and Sumikawa, K. Nicotine-induced switch in the nicotinic cholinergic mechanisms of facilitation of long-term potentiation induction. *Eur. J. Neurosci.*, 22(4): 845-860, 2005.

Yang, X., Criswell, H.E., Simson, P., Moy, S. and Breese, G.R. Evidence for a selective effect of ethanol on NMDA responses: ethanol affects a subtype of the ifenprodil-sensitive NMDA receptors. *J. Pharm. Exp. Ther.,* 278(1): 114-124, 1996.

INDEX

A

B

C

D

E

F

G

N

O

P

Q

R

S

T

U

V

W

X

Y

Z